Lloyd's Modern Poultry Book
A Poultry Guide and Directory

by W.B. Lloyd

with an introduction by Jackson Chambers

The World's Largest Selection of Vintage Poultry Books

www.VintagePoultry.com

Introduction

I am pleased to present yet another title on Poultry.

The work is in the Public Domain and is re-printed here in accordance with Federal Laws.

As with all reprinted books of this age that are intended to perfectly reproduce the original edition, considerable pains and effort had to be undertaken to correct fading and sometimes outright damage to existing proofs of this title. At times, this task is quite monumental, requiring an almost total "rebuilding" of some pages from digital proofs of multiple copies. Despite this, imperfections still sometimes exist in the final proof and may detract from the visual appearance of the text.

I hope you enjoy reading this book as much as I enjoyed making it available to readers again.

Jackson Chambers

Self Reliance Books

Get more historic titles on animal and stock breeding, gardening and old fashioned skills by visiting us at:

http://selfreliancebooks.blogspot.com/

2

PREFACE.

When I was eleven years old my father gave me a small flock of hens. Since then, except while in the army, and part of the time while away from home attending school, I have been personally and financially interested in poultry. I have raised hens, ducks and turkeys; raised them in village and upon the farm, in small and in large flocks; raised them for the eggs they would produce, to sell as spring chickens, and to turn off in the fall.

When in 1888 I became Agricultural Editor of the FARM, FIELD AND FIRESIDE, the Poultry Department fell to my care, and all sorts of questions about poultry came, have been coming, and still come to me. In the following pages I try to answer these questions.

W. B. LLOYD.

GLEN ELLYN, ILL., January, 1894.

INTRODUCTION.

THE dictionary definition of poultry is, "Domestic fowls collectively; those birds which are ordinarily kept in a state of domestication for their flesh, eggs or feathers, as the domestic hen, turkeys, guinea-fowl, geese and ducks."

There are so many things connected with poultry, besides the birds themselves, that a book on the subject may naturally treat of their care, their houses and other matters pertaining to them.

As this book is written to answer questions that come to my desk every day, the reader may consider each chapter or subject preceded by a question. The following are samples: "What are capons?" "Please describe the Redcaps." "Which is the best breed for layers?" "Please give description of a cheap hen-house." "What will cure bumble-foot?" "Where can I get Leghorn eggs?" "Where is the Reliable Incubator made?" Is there any profit in raising chickens on a large scale?" "What is the standard weight of a Toulouse gander?" "How should poultry be dressed for the Boston market?"

In answering these questions I have given my own experience, the experience of others who have been particularly successful in special lines of poultry raising, or the judgment of those who have been long in the business. In the list of breeders and of dealers in incubators and other supplies, only those are named whom I believe to be reliable and worthy of patronage.

CONTENTS.

DEFINITIONS.

[For the following Definitions and Nomenclature from the American Standard of Perfection we are indebted to the courtesy of the American Poultry Association. Everyone who desires to know the standard of points in all recognized breeds of poultry ought to own a copy of the Standard of Perfection. It can be had postpaid for one dollar.]

Barring.—Marks or stripes across the feather at right angles, or nearly so, to its length.

Beard.—A bunch of feathers under the throat of some breeds of chickens, such as Houdans and Polish.

Breed.—Any race of fowls having distinctive characteristics in common. Breed is a broader term than variety and may include several varieties, as the Plymouth Rock has Single-combed Barred, Pea-combed Barred and White as varieties of the breed.

Brood.—The family of chicks belonging to a single mother.

Broody.—Desiring to sit or incubate.

Cape.—The feathers under and at the base of the hackle, shaped like a cape. This term is most frequently applied to the Light Brahma, whose cape is composed of black and white feathers.

Carriage.—The attitude or "style" of a bird.

Carunculated.—Covered with small fleshy protuberances, as on the head and neck of a turkey-cock.

Chick.—A newly-hatched fowl.

Chicken.—A term indefinitely applied to any age under one year old.

Clutch.—A term applied both to the batch of eggs sat upon by a fowl, and to the brood of chickens hatched therefrom.

Cock.—A male fowl over one year old.

Cockerel.—A male fowl under one year old.

Comb.—The fleshy protuberance growing on the top of a fowl's head. The four chief varieties of comb are single, rose, pea and leaf; all others being modifications of and properly classed with them.

Condition.—The state of the fowl as regards health and beauty of plumage.

Crest.—A tuft of feathers on the head, of the same significance as top-knot.

Crop.—The receptacle in which a fowl's food is stored before passing into the gizzard for digestion.

Cushion.—The mass of feathers over the rump of a hen, covering the tail—chiefly developed in Cochins.

Dubbing.—Cutting off the comb, wattles and ear-lobes so as to leave the head smooth and clean.

Duck-foot.—The carrying of the hinder toe forward.

Ear-lobes.—The folds of bare skin hanging just below the ears—by many called deaf-ears. They vary in color, being red, white, blue and cream-colored.

Face.—The bare skin around the eye.

Flights.—The primary feathers of the wing used in flying, but tucked under the wings out of sight, when at rest.

Fluff.—Soft, downy feathers about the thighs and covering the posterior part of the bird, chiefly developed in Asiatics.

Furnished.—When a cockerel has obtained his full tail, comb, hackles, etc., he is said to be furnished.

Gills.—The same as wattles, which see.

Hackle.—The neck plumage of both sexes.

Hackles.—The peculiar, long, narrow feathers on the neck of fowls.

Henny or *Hen-feathered.*—The plumage of a cock resembling that of a hen from the absence of hackles and sickle-feathers, and in plumage generally.

Hock.—The joint between the thigh and shank.

Keel.—The breast-bone, so called from its resemblance to the keel of a boat.

Knock-kneed.—A term used to express an inward turning of the hocks by which they are brought together while the legs extend outward and are well spread at the feet.

Leaf-comb.—The two-pronged, V-shaped comb, such as is

seen in crested breeds, so called from the fancied resemblance to the open leaves of a book.

Leg.—In a living fowl this is the scaly part usually denominated the shank; in a dressed fowl, it refers to the joint above.

Leg-feathers.—Feathers growing upon the outer sides of the shanks, as in Asiatics.

Mossy.—Confused or indistinct marking in the plumage.

PEA-COMB.

Nub-comb.—An irregular pea-comb, but lacking in the true triple character, the longitudinal depressions or channels being grown up. It approaches in character to a rose-comb, but is properly classed as a pea-comb, as it is produced only by pea-combed varieties.

Pea-comb.—A triple comb, resembling three small single combs joined together at base and rear, lower and narrower at front and rear than center, and distinctly divided, the largest and highest in the middle, each part slightly and evenly serrated, as may be seen in the illustration above.

Penciling.—Small markings or stripes over a feather. These may run straight across, when they are frequently called bars, or follow the outline of the feather, taking a crescentic form.

Poult.—A young turkey.

Primaries.—The flight-feathers of the wings, hidden when the wing is closed, being tucked under the visible wing, composed of the secondary feathers. Usually the primaries contain the deepest color belonging to the fowl, except the tail, and great importance is attached to their color by breeders.

Profile.—A direct side view or illustration of a fowl.

Pullet.—A female fowl under one year old.

Rooster.—A term for a cock or cockerel.

Rose-comb.—A low, thick, solid comb, the upper surface

of which is usually corrugated or covered with small points. It usually terminates in a well-developed spike, which may turn upward as in the Hamburgs, remain nearly level as in the Rose-comb Leghorns, or turn downward as in the Wyandottes. In some varieties the spike is wholly wanting, or but slightly developed.

Saddle.—The posterior part of the back, reaching to the tail in a cock, and answering to the cushion in a hen—cushion, however, being restricted to a very considerable development, as in Cochins, while "saddle" may be applied to any breed.

Secondaries.—The quill-feathers of the wings, which are visible when the wing is folded.

Self-color.—A uniform tint over the feather, or a uniform hue to the plumage, in the latter sense being applied to all solid-colored varieties, such as white, black and buff.

Shaft.—The stem or quill part of a feather.

Shank.—The lower and scaly joint of the leg.

Sickles.—The long, curved feathers of a cock's tail, properly applied only to the top pair, but sometimes used for one or two pairs besides.

Single Comb.—An upright comb, varying in size and depth of serration, rising from the beak and generally extending back of the head for some distance, and consisting of a single thin, fleshy mass.

Spangling.—The marking produced by a large spot or splash on each feather, differing from that of the ground color.

Spur.—The sharp defensive weapon of the cock, growing from the inner side of the shank.

Squirrel-tailed.—The tail projecting over the back in front of a perpendicular line drawn from the roots of the tail.

Stag.—A term used for a young cock, chiefly employed by Game fanciers.

Station.—An ideal standard for Games, embodied in style and symmetry.

Strain.—A race of fowls that has been carefully bred by

one breeder, or his successor, for a number of years, and has acquired an individual character of its own.

Surface-color.—The color of the plumage or feather which lies upon the surface of a fowl when in a normal position and condition.

Symmetry.—Perfection of proportion; harmony of all the parts of a fowl, taken as a whole, and must be typical of the variety it represents.

Tail-coverts.—The soft, glossy, curved feathers at the sides of the lower part of the tail, usually of the same color as the tail.

Tail-feathers.—The straight and stiff feathers of the tail only; the top pair are sometimes slightly curved, but they are, generally, nearly if not quite straight and are contained inside the sickles and tail-coverts.

Thighs.—The joints above the shanks, the same as the drum-sticks in dressed fowls.

Top-knot.—The same as crest.

Trio.—A cock or cockerel and two hens or pullets.

Under-color—The color of plumage not exposed when the fowl is in a normal condition and position, and is seen when the surface has been lifted. It is manifested chiefly in the down seen about the roots of the feathers.

Variety.—A term used to denominate fowls possessing common characteristics, less wide in its application than breed, which see.

Venetianed.—Lapping over like the Venetian blinds used in houses. This term is frequently applied to the lapping of the tail-feathers.

Vulture-hock.—Stiff, projecting feathers at the hock-joint. The feathers must be both stiff and projecting to be thus truly called and condemned. See illustration on opposite page.

Wattles.—The red, depending structures at each side of the base of the beak, chiefly developed in males.

Web.—The web of a feather is the flat or plume portion; of the feet, the flat skin between the toes; of the wings, the triangular skin seen where the wings are extended.

Wing-bar.—A line of dark color across the middle of the wings, caused by the color or marking of the feathers known as the lower wing-coverts.

Wing-bay.—The triangular section of the wing, below the wing-bar, formed by the exposed portion of the secondaries

VULTURE-HOCK.

when the wing is folded. Used chiefly in reference to Game fowls.

Wing-bow.—The upper or shoulder part of the wing.

Wing-butts.—The ends of the primaries, also called wing-points.

Wing-coverts.—The broad feathers covering the roots of the secondary quills.

Wing-fronts.—The front edge of the wing at the shoulder. This section of the wings is sometimes erroneously called wing-butts, but the latter term should be applied only to the ends of the primaries to avoid confusion.

NOMENCLATURE.

1. Comb.
2. Face.
3. Wattles.
4. Ear-lobes.
5. Hackle.
6. Breast,
7. Back.
8. Saddle.
9. Saddle-feathers.
10. Sickles.
11. Tail-coverts.

12. Main Tail-feathers.
13. Wing-bow.
14. Wing-coverts, forming wing-bar
15. Secondaries.
16. Primaries or flight-feathers.
17. Point of Breast Bone.
18. Thighs.
19. Hocks.
20. Shanks or Legs.
21. Spur.
22. Toes or Claws.

Lloyd's Modern Poultry Book.

CHAPTER I.

THE BREEDS OF POULTRY.

Including all varieties of Games and Bantams, there are several hundred breeds of fowls, and to give a brief description of each would alone make a book of fair size, so we shall be content with giving a short account of some of the leading breeds, some of the newer ones and some characteristics or features of others.

The poultry of this country deemed worthy of recognition by the American Poultry Association is divided into thirteen classes. Ten of these embrace the domestic fowls, while there is one class each for turkeys, ducks and geese. The names given the classes of domestic fowls are largely derived from the countries where the different breeds included in them had their origin.

Included in these thirteen classes are thirty-three breeds of fowls, one of turkeys, eight of ducks and six of geese. These breeds are subdivided into varieties until the number exceeds a hundred.

The American class of fowls includes the American Dominique, the Black, the Mottled and the White Java, the Jersey Blue, the Barred, the Pea-comb Barred, the Buff, and the White Plymouth Rocks, the Buff, the Golden, the Silver and the White Wyandottes.

In the Asiatic class are the Light and the Dark Brahmas, the Black, the Buff, the Partridge and the White Cochins, and the Black Langshans.

Bantams are classed as the Booted White, the Game, the Black, the White and the Partridge Cochin, the Malay, the Black and the White Japanese, the Buff Pekin (or Cochin), the White-crested White Polish, the Black Rose-combed, the White Rose-combed, the Golden Sebright, and the Silver Sebright.

The White, Silver Gray, and Colored Dorkings are classed as English, and the Mottled Houdan, the Black Crevecœur, and the Black LaFleche as French.

Among Games are the Black, the Black-breasted Red and the Brown-red, the Golden Duckwing, the Silver Duckwing, the Red Pyle, the White, and the Black Sumatra. There are the same varieties of Game Bantams as Games.

In the Hamburg group are Black, Golden-Penciled, Golden-Spangled, Silver-Penciled, Silver-Spangled and White Hamburgs and the Redcaps.

The Mediterranean class includes the Black, the Brown, the Buff, the Rose-Comb Brown, the Dominique, the White and the Rose-Comb White Leghorns, the Black and the White Minorcas, the Blue Andalusians and the Black Spanish.

The various Polish varieties, including the Bearded Golden, the Bearded Silver, the Bearded White, the Buff Laced, the White-Crested Black, the Golden, the Silver and the White, are in a class by themselves.

Besides the above groups there are the Black Russians, the White Silkies, the Frizzles, the Rumpless, and the White Sultans.

Turkeys are distinguished as Black, Bronze, Buff, Narragansett, Slate, and White.

In ducks there are Pekin, Aylesbury and Crested that are white; Cayuga and East India, black; Muscovy, colored and white; Call, gray and white, and the Rouen that is colored.

Geese are classed as African, Canada, Chinese, Egyptian, Embden, and Toulouse. The African, Canada and Toulouse are gray, the Embden white, and the Egyptian colored. There are both brown and white Chinese.

UNRECOGNIZED BREEDS.

New breeds, or new varieties of old breeds, are constantly coming forward claiming recognition. Among those of merit not yet recognized by the American Poultry Association are the Anconas, Argonauts, Orpingtons, Sherwoods, Violettes and White Wonders. The Curassow, Guineas, "Pet Stock," Pigeons, Pheasants and Swans are not recognized as poultry.

PLYMOUTH ROCKS.

About forty-five years ago one Dr. Bennett thus described a breed of fowls to which he gave the name Plymouth Rock: "I have given this name to a very extra breed of fowls, which I produced by crossing a Cochin-China cockerel with a hen that was herself a cross between the fawn-colored Dorking, the Great Malay, and the Wild Indian. Her weight is six pounds, seven ounces. The Plymouth Rock fowl, then, is in reality one-half Cochin-China, one-fourth fawn-colored Dorking, one-eighth Great Malay and one-eighth Wild Indian. Their plumage is rich and variegated, the cocks usually red and speckled and the pullets darkish brown. They are very fine fleshed, and early fit for the table. Their legs are very large, and usually blue or green, but occasionally yellow or white, generally having five toes upon each foot; some have the legs feathered but this is not usual."

This "very extra breed" of fowls that bred legs yellow, white, blue or green, feathered or clean. five-toed or four-toed, was too "extra" to last long in this matter-of-fact world, and it is doubtful if our modern Barred Plymouth Rock with its beautiful dark or light steel gray dress is in any way very closely related or connected to its namesake of forty-five years ago.

The Plymouth Rock is our pet and we can no better state our appreciation of it than to use the words of a noted breeder of the breed. He says:

The Barred Plymouth Rock has been termed America's Idol. There is no other variety—the product of American skill in breeding—that we can put on the markets of the world with so much pride, and no other is received from our shores by foreign fanciers with such favor. The strongest proof of the superiority of this breed is that it has successfully stood the booms of a score of new varieties—has met and vanquished each one — and still lives. Other breeds have their booms, but the Plymouth Rock keeps on in its steady course, winning greater and greater popularity with each succeeding year. It has never had a boom in its history. Its favor has been won by merit, and by merit it retains what

PAIR OF BARRED PLYMOUTH ROCKS

PEN OF WHITE PLYMOUTH ROCKS—SEE PAGE 18.

it has won. It is the farmer's delight, the poulterer's "stand-by," and the villager's best friend, as it can be depended on to yield a generous supply of eggs and savory flesh. For meat, size, laying qualities, vigor, quick growing, and attractiveness combined there is no breed that will not suffer by comparison. This difficult union of qualities makes the Barred Plymouth Rocks continual favorites. For the market they are one of the best, being large, with plump bodies and full breast, with clean yellow legs and skin. For these reasons and many others they always command a high price. For table fowls they have but few equals, being sweet, juicy, fine-grained, tender and delicate. As layers they are considered above the average. Many breeds stop laying during the winter, but the steady-going Plymouth Rocks bid defiance to the season, provided their house is a warm one and they are plentifully supplied with food convertible into eggs. They always mature early and are splendid foragers, fast growers, and compactly built. Their heavy weight and short wing feathers prevent them from flying high, hence they are easily confined. The hens are the most patient of brooders and the best of mothers, and so determined are they to bring up a family that they often adopt a stray chicken. They are very hardy and healthy, thriving well in any weather. In looks the Plymouth Rocks may not take the lead, but just here the old adage applies, "Handsome is that handsome does." Their plain, Quaker-like attire is a suitable every-day work dress, and even those farmers who have an inborn dislike to "fancy chickens" cannot but admit that the pure-bred Plymouth Rocks are far ahead of any cross for farm stock.

In plumage they are a bluish gray, each feather distinctly penciled across, the bars of a darker color, and the more distinct the bar the better and more valuable the fowl. The plumage of the hen is much darker than that of the cock. They have straight, rather small size, single combs with five or six serrations, bright yellow beaks, red ear-lobes, and bright red wattles of medium size. The standard weight of the cock is nine and one-half pounds, cockerel eight, hen seven and one-half, and pullet six and one-half.

WHITE PLYMOUTH ROCKS.

Except in plumage, the White Plymouth Rock is the counterpart of the Barred variety, from which it is a sport. As general-purpose fowls, they rank with the best. They have no need to be ashamed of their record as egg-producers, nor has their owner. They lay especially well in winter. Their eggs are large, of good color, and excellent quality. As dressed poultry they are unsurpassed. The absence of the dark pin-feathers, and the rich, yellow skin, covering a plump, meaty body, make an attractive market bird. They are desirable fowls for table use, the flesh being white, tender, and fine-flavored. They have very large frames, and are somewhat slow in reaching maturity. The chicks are hardy, vigorous, and very tenacious of life. Add to these qualities the fact that they are very handsome and pleasing in appearance, there seems little else to be desired.

BUFF LEGHORNS.

This variety of Leghorns is claiming the attention of fanciers and is gaining friends right along, though there is some opposition to them, as there is to every new breed or variety of established breeds. There are but few fanciers in England or America who are pushing these new beauties— for they are beauties. One of the largest breeders of them says:

"In the Buff Leghorns we have usefulness and beauty combined; they are a grand table fowl and just the thing for the 'broiler man;' they are hardy and easily raised; mature very early; lay when four months old. They are wonderful layers, and are said to lay more eggs than any other variety of Leghorns by those who have tried all the varieties; lay a larger egg than the White or Brown; they bear confinement well, and are free from diseases to a great extent. Some find fault with them because they do not breed true. We claim they breed fully as true as many of the older breeds, such as the Golden and Silver Wyandottes and others, and are improving very fast in this respect.

"The Buff Leghorn is the same in comb, lobes, color of legs and shape as the White or Brown, only differing in color, which is a rich, even buff throughout; they are some larger than the other varieties of Leghorns. Many new breeds have come up in the last ten years, their friends claiming them to be superior to the Leghorns as layers, but after trial they have been found wanting, and the Leghorns to-day stand pre-eminently above all other breeds as egg producers, and no fowls have as many friends as the Leghorns of the different varieties.

"We have the past season stocked several farms with 'off' colored females. The owners of these farms have heretofore kept the Brown Leghorns, and have had the reputation of getting more eggs than any farmers of the neighborhood. They now report the Buff Leghorns being their best layers by odds. As chicks we have never seen their equal for hardiness. In quite a number of cases every egg hatched and every chick was raised. As fast growers they have no equals. We would advise all lovers of the Leghorn to give the Buff a trial."

WHITE-CRESTED BLACK POLISH.

The White-Crested Black Polish is a very stylish and ornamental bird, the color of the crest, and that of the rest of the plumage, making a most striking and beautiful contrast. With the exception of the white crest, the plumage is of a rich, glossy, greenish black; ear-lobes white, and legs dark. The cock will average about five and one-half pounds in weight, and the hens about four and one-half pounds. They are of a very tame disposition, like all of the Polish varieties. As egg producers, they are very profitable, being non-sitters, and most perpetual layers, even in seasons when common hens quit. Of medium size, plump and neat when dressed, with excellent flesh, they excel as a small table and market bird.

They may be considered a hardy fowl, standing our severe winters well, and, when common sense provisions are made for their comfort, coming out in good condition in the spring. The White-Crested Black Polish are recommended to those who wish both a useful and ornamental fowl for their park or lawn, as no variety, perhaps, is more admired for its attractive appearance and oddity. Among all the

varying changes in the poultry fancy, this old variety still
manages to hold a place. It is sometimes a matter of surprise
that it has not been neglected and become extinct in the rush
after newer and larger varieties. But, although in the past
sixty years they have been quite scarce, still some one has clung
to them like an old love, and so kept them from being entirely blotted out. Among all the great variety of fancy
poultry known to the fraternity, few, if any, can ante-date
the Polish, and it is a strong argument in favor of in-breeding that they have existed so long, and retained all their peculiar characteristics, for in-breeding must have, of necessity,
been considerably practiced to have kept them from dying
out. Long before the Cochins, Brahmas, or Leghorns were
known among us, the White-Crested Black Polish was bred,
and, while it was known as a layer of large, white eggs, still
it was its glossy black plumage and its large white crest
that proved its greatest attraction. Its origin is, I believe,
wrapped in obscurity, and, while it has always been known
as the Poland or Polish fowl, it is said to be a misnomer, as
its origin can not be traced to that far-away country.

THE ARGONAUT.

H. S. Babcock, the originator of this breed, says: The
Argonaut is a general-purpose fowl. In its making, this object was kept steadily in view, and the attempt was made to
unite great laying and great table qualities in a union as near
perfect as possible. These qualities are antagonistic, and
the gaining of one means, to some extent, the loss of the
other; but, as our American breeds prove, it is possible to
unite very good laying and very good table qualities in one
fowl. The Argonaut is another proof of the possibility of
making this very desirable combination,

The Argonaut is also an ornamental fowl, for it has a
beautiful shape and beautiful color. The shape, while sug-
gestive of that of the Game and the Plymouth Rock, is one
that is peculiar to the breed, and combines solidity with
gracefulness in more than an ordinary way.

The color is buff—one of the richest and most practical
colors that a fowl can possess, for its beauty is acknowledged
by everyone, while its practical character is easily seen from

the following considerations: It looks well in all situations, when fresh and new, or even when faded; it shows soiling as little as any color well can; it is the color that best goes with yellow shanks and yellow skin, a most desirable thing for the American market; it is a color that dresses handsomely, for the pin-feathers are very inconspicuous, there being but one color superior to buff in this respect—white, but that has other objections, which puts buff in the lead. A color so beautiful and so useful as buff, upon a fowl of its solid yet graceful shape, gives it strong claims to be regarded as an ornamental fowl.

The Argonaut is unique in one respect—it is the only buff pea-combed breed in the world; and it breeds very good pea-combs, too. For a practical fowl, I regard this as a valuable characteristic, as it removes the comb as far as possible from the danger of frost.

It may be said that, as originator of the Argonaut, I am a prejudiced witness. I admit that I am, but this very fact has kept me from praising the fowl according to what I think are its deserts, and this article, instead of being overdrawn, falls short of doing the breed justice. Of its many admirable qualities I have given but a faint hint, and upon its history I have been silent.

If I did not regard it as a valuable and needed addition to the useful breeds in this country—valuable because it unites good laying with fine table qualities, and needed because there are thousands of admirers of this color who can not otherwise obtain it upon a general-purpose fowl with clean, yellow legs—I certainly should not have spent the time and money in its production which I have, and I should not breed it, even if I had been foolish enough to produce it, unless I had a large faith in its permanent value to the poultry stock of the country.

WHITE COCHINS

Are magnificent fowls, and worthy of a more general cultivation among American farmers. Being pure white, they are much easier to breed true to color than a party-colored variety. The plumage is white; wattles and ear-lobes brilliant red; comb red and single; legs well feathered to outer toe. These fowls should have a shaded run in summer, or the glare of the sun day after day will give their plumage a yellowish appearance. They are among the hardiest of fowls, feather and mature extra early, and are good winter layers. Cocks weigh eleven pounds, hens nine pounds.

PARTRIDGE COCHINS.

The plumage of the Partridge Cochin is very rich and elegant, resembling a partridge to some extent. Their gait is slow but the carriage quiet and dignified. The Cochin is not adapted to any particular climate or section of country but thrives equally in Canada or Florida, on the Atlantic or Pacific Coast. In the winter they are good layers. They are very large and are a profitable cross for increasing the size, hardiness and early maturing of barnyard fowls, for which purpose they may be profitably used by farmers. They may be confined by a low fence, three feet being ample height. Cochins will thrive well in the smallest yards, and under such unfavorable circumstances as to preclude the successful rearing of other fancy breeds.

BUFF COCHINS

Are growing more popular every day, and deservedly so. Too much cannot be said about the excellence of this variety. There is no breed that is more thoroughbred or more admired by the fraternity in general. Considering them from a standpoint of utility, they fill the bill as well as any of the Asiatics. The secret of their great popularity is because of their superiority. The American Buff Cochin Club is doing much to advance them. and the secretary of the club says he will have them at the head of the list before long. Uniting great size, hardiness, and much more than average egg production in the same fowl, Buff Cochins have a strong claim upon that class of people that desire to make money out of a few hens; while their majestic form and exquisite color make them much admired by breeders of thoroughbred fowls. The plumage is a clear, beautiful shade throughout, the neck, saddle, and tail-covers being of a darker and richer shade in the cock.

The great fault which beset this breed originally (the

constant appearance of dark feathers) has been overcome at last; by careful mating and breeding they have reached that state of perfection where the fancier has comparatively little trouble in breeding them to feather. Plentifully covered with long, soft feathers, Buff Cochins look larger than they really are. The abundant covering of downlike feathers renders them less sensitive to extreme changes during the winter months, hence they are reliable egg producers during cold weather. The Buff Cochin is a very domestic fowl, does not fly high nor wander far, is a persistent and faithful sitter, is a mother that does not weary of maternal duties, and is faithful to her charge.

BLACK HAMBURGS.

All Hamburgs possess the same general characteristics; stylish and active in carriage, slender, rather short, blue or slaty-blue legs, with deep red rose-combs and close-fitting, pure white ear-lobes. They require free range, and are then easily kept, as they are excellent foragers. They will lay upward of 200 eggs in a year. Their eggs are not so large as those of the Leghorns; yet, as long as eggs are sold by the dozen, this makes little material difference in supplying the market. Mr. A. Beldon says of their early maturity, he has found that pullets of the penciled varieties lay at five months; the spangled not quite so early.

The Blacks are the largest of the Hamburgs and lay the largest eggs. They are also considered the most hardy. A great fault with many Black Hamburgs is a tendency to white on the face. This disqualifies pullets and cockerels. The face must be one rich, deep red, like the wattles, contrasting strikingly with the pure white ear-lobes.

GOLDEN POLISH.

There are two varieties of the Golden Polish—the bearded and the unbearded—the former being more attractive, as the beard is a fitting counterpart to the beautiful crest of the head. Within the last few years the bearded variety has grown so in popularity that it will, in course of time, entirely supplant the non-bearded. The general color of their plumage is a rich golden bay, each feather marked with black, in the form of a spangle or lacing, the marking increasing in size with the size of the feather. Like all spangled or penciled fowls, the marking of this variety is very beautiful. Their characteristics are like all other Polish. They are quite docile in dispositson, and fond of being petted. The hens

GOLDEN POLISH COCK—SEE PAGE 23.

are of medium size, good layers, and non-sitters. Their eggs are of pure white color and ordinary size.

THE GUINEA-FOWL.

Although this bird is not recognized among fanciers as poultry, there are many who consider it worthy a place on the farm. The following descriptive sketch, written by W. Willis Harris, appeared in a recent number of the Canadian Poultry Review:

There are several varieties of this bird, which is a native of Africa. The two varieties most generally bred under domestication are the speckled or pearl, and the white, the speckled being the much more common variety of the two. The unpopularity of the Guinea-fowl is chiefly due to its wandering habits, the difficulty of finding its eggs, which are laid in very secluded places, and the unpleasant noise it gives vent to, very much resembling the grating of a cartwheel; but the latter has its advantages, making a flock as valuable at night as a first-class watch-dog.

As game it has proved a failure, for when turned down in the coverts it drives away the pheasants, and will not rise to the gun, but will run before the dogs at a marvelously rapid speed. In the poultry-yard it is very spiteful (especially the cock) to young chicks, and is, generally speaking, of a very pugnacious disposition. But in spite of these dis-

PAIR OF WHITE COCHINS—SEE PAGE 21.

advantages, as a semi-domesticated bird, it is very profitable upon a farm or anywhere where it can have free range and plenty of liberty, clearing the ground of myriads of insect life, and being a small feeder in comparison with ordinary poultry.

From March to October the female lays a great number of eggs of a speckled cream-color, with hard shells, averaging during the season 150. Their nests are very secluded, and are generally made in the center of a thick hedge, in the midst of shrubbery, or in the depth of a copse. As they are very cunning in the selection of their nesting places, their eggs are somewhat difficult to find, but can best be discovered by watching any suspected spot, when the cock will be seen keeping guard whilst his mate is laying. The nest discovered, the eggs should be removed daily, two or three being left, or dummies substituted, otherwise the hen will desert the nest and make another in a still more secluded place. Several hens deposit their eggs in one nest, and it is therefore no uncommon thing to find twenty to fifty in a batch.

It is advisable to start keeping Guinea-fowls by either purchasing eggs and hatching them under domestic hens, or

PAIR OF PARTRIDGE COCHINS—SEE PAGE 22.

procuring them when young, when they are more likely to localize themselves to their owner's wish than if purchased as older birds. If adult birds be purchased, they will require boxing up for three weeks or a month and feeding carefully to tame them, otherwise they are liable to wander off at their own sweet will, possibly never to return.

It is seldom the eggs are infertile, and they should be set in rather a damp nest, the eggs requiring more moisture than

those of the common fowl. It is better to set clutches of fifteen to eighteen eggs under ordinary hens (half-breed Game
preferred), as the Guinea-hen seldom sits until the latter end
of August, which is too late in the season for the young birds

BUFF COCHIN—SEE PAGE 22.

to thrive, as they have not the stamina to withstand the
early frost and autumnal wet. The period of incubation is
twenty-six to twenty-eight days, and if the eggs be fresh the
chicks hatch out strong, and are of a brown color, striped
more than spotted, with bright red legs.

For the first three or four weeks it is absolutely necessary to fix a wire run in front of the coop in which the hen
and her chicks are penned, until the young ones have become

used to the call of the mother, or they will quickly ramble away, which they do far from slowly and the major
portion of the brood will be lost.   They are somewhat
delicate when young, but not so difficult to rear as turkeys
or pheasants, requiring to be similarly treated and fed.
The grass on which they are fed should be kept closely
mown; insects and animal food or its substitute "crissel,"
or bullocks' liver chopped fine, is absolutely essential to
successfully rearing the Guinea-fowl.   The chicks should
be fed for the first few weeks regularly five or six times a day;
biscuit meal makes an excellent staple food, varied with oatmeal and small corn at night.   At five or six weeks old they
commence to put on their adult plumage, and may be allowed full liberty with the hen.   At the age of three months
they develop the wattles and horny crests on the top of
their heads.   The sexes are somewhat difficult to distinguish
but the male is the larger of the two, and the wattles and
horn of the cock are larger than those of the hen.   It is only
the female that cries "come back, come back;" the cocks
when running after the hens arch their backs, and run in a
mincing way as if on tiptoe.

The Guinea-fowl in a wild state is monogamous; but under domestication some state they have run one cock with
three or four hens successfully, but I think it would be safer
to run in pairs.   They are gregarious, and a flock reared together will always continue to run in company and roost in
the same tree.   It is seldom they can be induced to roost in
ordinary poultry-houses or to lay in nests provided for them,
preferring the semi-wild state, wandering with sweet liberty
through copse and meadow; and though natives of a hot, arid
climate, braving the roughest of weather, and not being
poisoned with the close atmosphere of artificial housing, they
are, when mature, practically free from disease.

The adult birds should be fed similarly to ordinary poultry, but require insect or animal food, which, if at liberty, they
will find for themselves.   It is also advisable to feed at regular hours, particularly at night-time, so as to induce them to
remain at home.   They come in season for the table from
Christmas to March; the poulterers have little difficulty in

disposing of them to their customers. Like game, they do not require fattening, but, similar to pheasants, they should be well hung previously to cooking.

There are but few shows in the country that provide classes for Guinea-fowls; they are generally exhibited in pairs. To prepare them for exhibition, they simply require to be penned for a fortnight or three weeks, to tame them, and their heads and legs washed, and rubbed over with a tiny piece of vaseline.

Mr. Willis throws out the following as a suggested standard in judging the

SPECKLED OR PEARL VARIETY.

Head—Broad, surmounted with a horny crest; wattles, a thick red, the freer from white patches the better.

Beak—Strong, curved; well-set in head; in color, pinkish horn.

Eyes—Bright, clear; color, steel gray.

Face—White, dotted with fine hairs.

Neck—Long, symmetrical curve; color, violet, purple, brown.

Back—Curving, rising from the neck to the center, and then descending in a graceful curve to the tail.

Breast—Broad and full.

Body—Deep through the center, with long keel.

Fluff—Short.

Wings—Close, tight-fitting, with few or no white feathers in flights.

Tail—Short.

Thighs—Short.

Shanks and Toes—Pink and black, the more evenly marked the better. Nails, light horn color.

Color of Plumage—Black, evenly marked with small white dots; the more evenly the better.

Disqualifications—Deformities of any kind. Any white or black feathers, except in the wing, the primaries of which may be white.

SCALE OF POINTS.

Symmetry..10

Size...25
Condition.......................................10
Head and Wattles................................10
Color...25
Color of Wing...................................10
Legs and Toes...................................10
                                              ——
                                              100

THE WHITE GUINEA

except in plumage, which is white throughout, is identical in appearance with the pearl variety. We have never raised any white ones, but one who has, says "they are fine layers of very rich flavored eggs. As layers they almost rival the Leghorns. The young are hardy, and easily raised if given the required attention.

"The flesh of the White Guinea, unlike that of the speckled variety, is very tender and toothsome. As a table fowl they come nearer the wild game birds than any of our domestic fowls. Their scarlet-trimmed heads and beautiful snow-white plumage attract many admirers. White Guineas grow rapidly, and for broilers excel young chickens, and farmers who want to raise something pretty as well as toothsome will do well to raise a flock of these pretty little birds."

LANGSHANS.

The Langshans are natives of Northern China, and were first sent from the Province of Langshan, by an employe of the British Government, to England. Some years ago they were brought to America, and fanciers generally consider them the best poultry acquisition we have from China. Langshans have straight red combs, somewhat larger than those of the Cochins. Their breast is full, broad and round, and carried well forward, being well meated, similar to the Dorkings. Their body is round and deep like the Brahmas. The universal color of the plumage is a rich metallic black. The tail is long, full-feathered, and of the same color as the body. The color of their legs is blue-black, with a purplish tint between the toes.

The good qualities claimed for the Langshans are: They

are hardy, withstanding readily even severest weather. They attain maturity quite as early as any of the larger breeds. They lay large, rich eggs all the year round, and are not inveterate sitters. Being of large size, with white flesh and skin, they make an excellent table fowl, more especially so on account of the delicacy of the flavor which the flesh possesses. Standard weight of cocks, nine and a half pounds;

PAIR LANGSHANS.

hens, seven pounds. They seem to combine all the characteristics that go to make up a practically useful fowl.

The accompanying faithful illustration will give a more accurate idea of them than an extended description. It will be observed that, apparently, they are more like the Black Cochin than any other breed with which we are familiar, but in reality they differ very essentially from them.

BLACK SUMATRAS.

Sumatras were introduced into the United States from the Island of Sumatra, and have been bred in limited numbers. The accompanying illustration shows their peculiar characteristics, small heads, pea-combs. pheasant-like tails, broad and sweeping low to the ground in full-plumed cocks. Their plumage is solid black, glossy and beautiful.   They are stout, active, rather small birds.   They are good layers and juicy table poultry.   In the latter respect they resemble wild

PAIR BLACK SUMATRAS.

game more than common poultry.   Although rather gamey in appearance they are not ranked as fighters.   When first hatched the chicks are nearly white and retain this color till the first moult.   After this, nearly all will assume the solid black with green luster.   In some, occasionally white feathers and in others a few red are seen.   These are faults and should be guarded against.   Their legs are clean, strong and of a dark leaden color.

HOUDANS.

The Houdan is to France what the Plymouth Rock is to America.   They derive their name from the little town of

Houdan, France. It is not positively known how they originated. Some believe they are a cross of the Black Polish
and the Dorking. They have the fifth toe of the Dorking
and a conformation of body between the two breeds. Their
plumage is made up entirely of black and white. They
have a large crest and beard which gives them a very pretty
appearance. Their legs are medium length, unfeathered,
pinkish white, mottled or shaded with black or lead color.

PAIR HOUDANS.

The fifth toe should be detached from the others and curve
upwards. They are a splendid table fowl, having fine, close-
grained meat in great plenty, and are considered one of the
best breeds for broilers.. They are excellent egg producers,
laying the largest egg of the non-sitting breeds. They
bear confinement well, but when given their liberty forage
well and will pick up their living equal to turkeys. The
standard weight of the cock is seven pounds, cockerel a
pound less, with the hen weighing the same and the pullet five
pounds.

CREVECŒURS.

The head of the Crevecœurs is quite small, and if standard bred they have a crest and beard. The crest should be
jet black, composed of feathers of the texture of the hackle

—large, round, close, well fitted on the crown—generally falling backward and rather lower on the sides of the head than over the beak; the comb is red, V-shaped (from which fact the name Crevecœur—literally heart-break—is derived), and of medium size; the eyes large and bright; the neck of medium length, and neatly carried a little over the back; breast deep and full, and carried well forward; back wide and straight; tail full and carried generally erect; thighs

PAIR CREVECŒURS.

short; legs slate color or black, and free from feathers. They have a watchful, upright and vivacious carriage, and are a brilliant black.

They rarely fly, always walk slowly, do not care to ramble. Are good layers, beginning a little later than the Brahmas and Cochins; but their eggs are very large. As a rule they do not sit.

They are a first-class table bird. A full grown cock weighs from eight to nine pounds, and a hen from seven. to eight.

Their beautiful black plumage, large crests, and two-horned combs make them conspicuous. Although much thought of in France, their native land, where they rank next to the Houdan for utility, they have not proved hardy in this country.

INDIAN GAME.

The Indian Game is rapidly taking first rank as a table fowl, the flesh being but little inferior to that of the turkey in flavor. It grows quickly and has a very large breast, thus making the best of broilers. Its table qualities are wonderful. The feathers are greenish-black with brown-crimson shafts; the legs are stout and of a yellow color. The bird is powerfully built and has a very broad body. It weighs as much as the Brahma, though it does not appear nearly so large. The hens are among the best of mothers, and are as good layers as the Plymouth Rocks.

LA FLECHE.

This breed, originating in France, has not as yet been extensively introduced in this country, it being the general belief that it is constitutionally too weak to bear the severe winters of the Northern States, though this may be due in large measure to in-breeding. Their ancestry is shrouded in mystery, but they are probably closely related to the Spanish, and, like them, do best in the mildest country.

The plumage is glossy black throughout, the shanks leaden black, the ear-lobes pure white like those of the Black Spanish. The comb is bright red and shaped like a pair of horns, pointing almost straight upward, with two small knobs in front of each horn. They are non-sitters and lay large white eggs of a rich flavor. The French consider their flesh very fine eating. The standard weight of the cock is eight and one-half pounds, and of the hen seven and one-half.

BLUE ANDALUSIANS.

This breed is claimed to have originated in the Province of Andalusia, Spain, from whence it takes its name. It is an older variety than the Minorca and resembles it in many respects. It is a hardy fowl, easily acclimated, and more docile than any of the other non-sitting breeds.

The plumage is blue and black, the comb single and rather large, and in the hen falls to one side, partly concealing the eye. It is a fair table bird, but is especially esteemed for its laying qualities, being considered superior to the Leghorns by the English; it also surpasses that breed in size.

SHERWOODS.

The Sherwoods were first brought to the notice of the general public in 1890, by W. Atlee Burpee & Co. They are a cross of White Georgia Games and Light Brahmas. They derive from their Brahma parentage a heavy body but are shorter in leg; from the Game parentage, fuller breasts. They

TYPICAL GUINEA FOWL—SEE PAGE 29.
Reproduced from Poultry World.

are very stylish birds and very majestic in carriage, with close, compact bodies. Their yellow bills, beautiful erect combs of medium size, bright red ear-lobes, white plumage, and yellow legs, slightly feathered to the outside toe, make them an attractive sight on the lawn. Their feathers are not fluffy, but are close, like the Indian Games. They endure the cold weather better than the Asiatics or other fowls of equal size. It is claimed the young chicks are hardier than any other breed in existence; damp weather seems to have little or no effect on them. They grow rapidly, mature early, and are ready for broilers at ten weeks. In fact, the chicks just out of the shell are almost double the weight of a Brahma at the same age. They are very careful and attentive mothers, yet gentle and tractable to handle. They lay equally as well as the Plymouth Rock and the eggs are of a very large size, white in color, fine flavor, and good quality. The birds are of good size, cocks weighing from ten to twelve pounds; hens eight to nine pounds. For table use they are unexcelled, the flesh being tender and delicious, partaking somewhat of the game flavor. They are considered one of the best general-purpose fowls.

WHITE WONDERS.

This breed, recently introduced into New England, seems to have been originally a cross between the White Wyandotte and Light Brahma. They closely resemble the former, but are larger and have feathered legs. They are superb layers, and highly esteemed as market fowls, having brought three cents above the highest price, as broilers, on the Boston market.

SILVER-GRAY DORKINGS.

The males have black breasts, silver-white hackles and saddles, black tail wing bows silvery white, wing coverts black, and black bodies. The females are fully as handsome. They have silver gray heads; hackles, silver-gray; breast, bright salmon; back gray. They have large square-shaped bodies, short legs, with five distinct toes on each foot. The comb is large and single, and the shanks are white or flesh-colored. The Dorking is the only distinctive English breed, and is prized by them for its large, tender, juicy breast meat. The White Dorking differs from the Silver-gray in having a rose comb and pure white plumage throughout.

The Colored variety has either a rose or single comb, and the plumage of the cock is black and white on neck and back, with black breast, body and tail; that of the hen is black on neck with gray edging, dark gray marked with black on the back, dark salmon marked with black on breast, body dark brown or black mixed with gray.

WHITE LEGHORNS.

This is the original of the Leghorn family, all others having been derived from it by crossing. Its pure white plumage and yellow legs make it a handsome bird. The comb of the cock is large and erect, that of the hen falls to one side. It is a non-sitter and is one of the best layers, the eggs

PAIR WHITE LEGHORNS.

PAIR BROWN LEGHORNS.

being as large as those of some of the heavy breeds. The Leg-horn is small but feathers quickly, making it valuable to the "broiler-men" for crossing with the Asiatics. The flesh is of a fine flavor, making a good fowl for family use. The pullets sometimes begin to lay when only four months old. There is

SILVER-SPANGLED HAMBURG—COCK.

a rose-combed variety which differs only in having the comb like that of the Hamburgs.

Black Leghorns in most points are similar to the single-comb White. Their shanks and toes are yellow or yellowish black and their plumage is a rich glossy black.

But for color of plumage the Dominique Leghorn in looks is identical with the single-comb White. The plumage throughout is a grayish white, each feather regularly crossed with parallel bars of blue-black, giving in effect a bluish tinge.

BROWN LEGHORNS.

This variety came from the White by a cross with the Black Red Game. They greatly resemble the former except in color, though they do not lay as large an egg. The plumage of the cock is reddish bay on head, dark red on back and

PAIR BLACK MINORCAS.

black on breast, with large, well curved tail of metallic black. The head and back of the hen are dark brown and breast salmon brown. There is also a rose-combed variety.

SILVER-SPANGLED HAMBURGS.

This is a very attractive breed, the plumage being white with large black spangles; its large, bright rose-comb adds much to its appearance. It is not a valuable market bird, on account of its small size, but in number of eggs it is hard to beat. It is a small eater and can be kept at little expense.

BLACK MINORCAS.

When the Black Minorca first made its appearance in America it was known as the Red-Faced Black Spanish. Black Minorcas are one of the handsomest breeds in the Spanish class. They are very stylish birds, majestic in carriage, with close, compact bodies, and low, well-set legs. The plumage is a beautiful, glossy black, shaded with the purple hue peculiar to some black birds. The face is coral red, with white ear-lobes. The legs are dark slate, or nearly black. They are exceptionally good layers, some claim equaling the Leghorns, and their eggs are much larger. A well-known English writer places the Minorcas first among all fowls as layers, and says: "They will lay from 200 to 225 eggs a year." One breeder of this variety has a record of 247 eggs from one Minorca hen in 365 days. The chicks mature rapidly, and are fit for broilers at from ten to twelve weeks. They are non-sitters, although they have been known to sit and rear their young. This, however, can be rarely depended upon. They possess a lively disposition, are very hardy, good foragers, and small eaters. Their useful and profitable qualities, combined with their handsome appearance, make them very popular. The mature cock should weigh eight pounds and the hen six and one-half.

White Minorcas, except in plumage, which is pure White, are exact counterparts of the Black variety.

THE LIGHT BRAHMA.

This is the largest of all the breeds, the standard weight of the cock being twelve pounds, and of the hen nine and a half. The hens are good mothers and lay large eggs. Like most of the other Asiatic breeds, they are quiet, docile and easily confined. They are generally kept for their flesh, as it is firm, juicy, of good flavor and great quantity, but they lack plumpness and have too much bone for broilers. The color is white with black hackles and tails. They have a small pea-comb and heavily feathered legs.

PAIR LIGHT BRAHMAS.

PAIR DARK BRAHMAS.

DARK BRAHMAS.

These birds closely resemble the Light Brahmas except in color. They also weigh a little less.

The cock has white hackles and saddies, each feather being streaked with black, the plumage on the breast is black, on the back and wings whitish. The general plumage of the hen is black with steel-gray pencilings.

VIOLETTES.

Little is known of this new and beautiful variety, and but few have seen them. They have the markings of the Golden Wyandotte, only instead of black they are a rich violet blue; this makes them very beautiful and attractive. All who have seen them pronounce them the prettiest thing out. The contrast between the blue and the gold makes them quite a novelty. In make-up, they have rose combs, with white or yellowish white ear-lobes, bright yellow legs, and in size, between the Wyandotte and the Leghorn, making them about a pound lighter than the Wyandotte. One of the originators of the breed says "there are two distinct strains, all being bred to the same ideal and helping each other, and in course of a year or two we will show the public what can be done by skill and careful breeding. We do not claim for them that they are better than other American varieties, but one thing sure, they will lay more eggs than the Wyandottes and still be a sitting breed. We are not claiming the earth, but when put on the market they will stand for themselves."

BLACK JAVAS.

The name might lead one to think that this breed came from the island bearing its name; on the contrary it was originated in Missouri about twenty years ago. It is hardier than the Plymouth Rock, and is equal to it in size and table qualities. The plumage is glossy black, the comb rather small and single, the legs black, the tail rather long and curved straight up and then back. Its yellow skin and plump body brings a ready sale on the "dressed poultry" market.

There are also White and Mottled (black and white) varieties, whose chief characteristics are the same as the Blacks.

WYANDOTTES.

Though one of the newest, this is one of the most popular of the American breeds. They are a hardy fowl, are one of the best layers of the sitting breeds and have plump yellow bodies and their flesh is of fine flavor. Their plumage and shape also make them favorites. In fact, taking all the desirable qualities into consideration, the Wyandotte is very properly called a "general-purpose fowl." Some go as far as to call it "the general-purpose fowl."

There are four varieties recognized by the American Poultry Association, the Silver, the Golden, the White, and the Buff; the Black, though not recognized, is becoming quite popular, and the Columbian, "latest born," is now claiming attention. The Silver was the original variety; early in the seventies, several leading breeders produced, from different crosses, breeds which closely resembled one another. These became known as American Sebrights; but before they were admitted to the "Standard," the cross between the Dark Brahma and the Silver Spangled Hamburg took the precedent and the name was changed to Wyandottes.

About this time a gentleman in Wisconsin perfected a cross between the original Wyandotte (now known as the Silver) and a "composite" fowl of his own breeding; this cross was called the Golden. It differs from the Silver only in having the body plumage reddish and golden bay instead of silvery white, the wings and tails of both varieties are black and the body feathers are striped through center with same color.

The White and the Black varieties are "sports" of the Silver, and are respectively solid white and black throughout.

The Buff is made up of many breeds and is claimed to have less Wyandotte characteristics than the others. It is of recent introduction to the public, not being known to one of the leading poultry journals of America two years ago.

The Columbian Wyandotte was first advertised as the Bra Wyandotte, a name probably intended to be descriptive of its appearance, for we are informed that it has the shape of the Wyandotte and the plumage of the Light Brahma. "Its originator, Mr. B. M. Briggs, assures us that it has not a drop of Brahma blood and that the variety was suggested by an accidental cross. What the cross was, he leaves us to guess as best we may; he volunteers no information upon the subject. With the White Wyandotte in the field and filling the demand for a white fowl of the size and the characteristics of the

SILVER WYANDOTTES.

Wyandotte, the chances of success for the Columbian seem to be heavily handicapped; should it succeed in spite of this fact, it will prove that the Light Brahma marking is one that suits the people and helps to account for the continued popularity of the favorite Asiatic fowl."

All the Wyandottes have rose combs and yellow legs. The shape much resembles that of the Dark Brahma but is more trim in appearance owing to its Hamburg blood. The mature cock should weigh at least eight and one-half pounds and the hen six and one-half pounds.

THE NAKED-NECKED FOWL.

This breed originated in Transylvania. The neck is bare nearly down to the breast, and is of a red color and

smooth.   There is a spot heavily feathered just on top of
the crop.   The chicks grow rapidly and feather more quickly than Hamburgs.   They are good table fowls and small eaters; the hens lay good-sized dark eggs, and plenty of them.   The
cock weighs about seven pounds, and the hen five.   The accom-

NAKED-NECKED FOWL.

panying illustration from Poultry (England) is a very life-
like representation—an accurate portrait of one shown at a
show in London.   In Germany, where this bird receives the
most attention, and where the best ones are kept, the plumage is black.   The Naked-Necked Fowls, while those who
have kept them have claimed for them many profitable qualities, have been mainly looked upon as curiosities, and have
been shown of several types and different styles of combs,
clean and feathered legs, and of various colors.

BANTAMS

are really nothing but dwarfs produced from the larger breeds, principally by crossing, late hatching and scant feeding. If hatched early and fed heavily their size will be increased.

Besides those recognized by American fanciers there are several varieties in process of development, and it is but a question of time, perseverance and skillful breeding when there will be a Bantam or dwarf of every standard breed and variety.

It is usually not the farmer with hundreds of acres of ground that keeps Bantams; but the city resident, with his little narrow back yard, may keep them with much pleasure. A few feet of ground and a dry, well ventilated dry-goods box for a coop, will do for a few of these fowls. They are both a useful and a fancy fowl. Prettiness is not all there is of them, for, in proportion to size of bird, there is no fowl that lays a larger egg than the Bantam. The small amount of food that they require and the goodly number of eggs that they lay are also in their favor.

BLACK AFRICAN BANTAMS

are the smallest of the hen tribe, weighing only ten to fifteen ounces. But they are as conceited as Peacocks. Their plumage is jet black and they have white ear-lobes and small rose-combs. The cocks have large sweeping tails which with their color and small size make them very attractive.

PEKIN OR COCHIN BANTAMS.

For some time there has been a variety of Bantams much admired for their quaint likeness to the well known Buff Cochin, and recently the Black, White and Partridge varieties have been originated. They will probably become popular among those who like an attractive fowl "built" on a small scale.

GOLDEN SEBRIGHT BANTAMS.

The Sebright Bantam is one of the hardiest varieties to breed to a high standard of perfection. In the opinion of some they are the most beautiful of all the Bantams and in many respects superior to any of the pigmies. Their fine plumage, of rich, golden yellow, laced with black, large rose-combs, with well-developed spikes, and blue legs, make them an object of beauty, admired by all who see them. It required years to produce these little fellows. The tail feathers of the cock are rarely laced with black; in many cases running all black, but if nicely tipped with black, will stand good in the closest of competition. He

PAIR OF GOLDEN SEBRIGHT BANTAMS.

should be perfectly hen-feathered throughout. Not only should his tail be free from sickle feathers, but the feathering on the neck and saddle should be like the hen. The hen-tailed cock is the correct type and a bird with pure white ear-lobes has obtained one of the rarest points on the Sebrights, as in most cases they run of a bluish tint, and in many cases they are all red.

The Silver Sebright only differs from the Golden in having the groundwork of the plumage silver white.

BOOTED WHITE BANTAMS.

These much resemble the Sebrights in shape and carriage, but their abundant hackle and saddle feathers and long sickles give them quite a different appearance. They are pure white throughout and have single combs. The thighs are furnished with long, stiff feathers, and the shanks are heavily feathered, hence the name "Booted."

WHITE-CRESTED WHITE POLISH BANTAMS.

This breed closely resembles the White-Crested Black Polish, except in size and color, although the comb may be either small and single or leaf (V-shaped). The plumage is

SILVER SEBRIGHT BANTAM.

pure white, and combined with the large crest makes them an attractive, ornamental fowls, for which they are especially adapted, being very domestic by nature, and readily submitting to petting.

JAPANESE BANTAMS

are among the most admired of the Bantams, yet it is probably their oddity rather than their beauty that gives them this distinction. They have very short legs, full breasts and low-carried wings, so that they often appear to be squatting when they are in reality standing. The cock has a large red comb and a very large tail, carried so erect that it often almost touches the back of his head. When mature they are quite hardy, though their combs are easily frost-bitten; as chicks they are rather delicate. Although they have been bred in a variety of colors, there are only three varieties recognized in America, the White, the Black and the Black-Tailed, which is pure white excepting the tail.

GAME BANTAMS.

There are seven varieties of these, namely: Black-Breasted Red, Brown Red, Golden, and Silver Duckwing, Red

PAIR OF JAPANESE BANTAMS.

Pyle, White and Black. They are the same in shape and color as the Games from which they are named, but the weight of the cock is only twenty-two ounces, while that of the hen is twenty.

BLACK-RED GAME BANTAMS.

Probably the most numerously bred Bantam is the Black-breasted Red. The brilliant pluming of the male and the slim, slick bodies of both sexes, with their upright carriage, make them very attractive. They are perfect miniature Games—exactly like the Black-Reds in color and style, only smaller. They are very handsome, saucy, independent and sprightly little pets. They lay abundantly, and, while their eggs are small, many prefer their flavor, when cooked, to those of larger breeds.

ROSE-COMBED BANTAMS.

These are just the opposite of Game Bantams in shape. The head is carried back over the body. The breast is carried well forward and the legs are set midway of the body. The tail is full and well expanded and the legs are short. The hens are good layers, producing eggs freely in the winter. The chicks mature rapidly and are very pugnacious at an early age, often killing one another when only six weeks old.

There are two varieties which differ little but in color. These are the White, pure white throughout, and the Black, lustrous black.

ODD VARIETIES OF BANTAMS.

Fly Fishers are an English variety with a slaty blue plumage like the Andalusian. The hackle feathers are used in making artificial flies for fishing, hence the name. They are probably a close relation to the Cuckoo Bantam.

The Silk Bantam is called a separate variety by some, but much resembles the Silkies, which are often no larger than a Bantam. Their plumage is very fine and much like silk.

Frizzled Bantams are another oddity, having recurved plumage. They are nothing but dwarf specimens of the Frizzled fowls, and are found only in England.

THE ORIGIN OF GAMES.

The origin of the Game cock is enveloped in considerable obscurity, for whilst many naturalists affirm that it is the reclaimed wild jungle fowl, as still found in India, many others who have given the subject much careful consideration and research, are of the opinion that our Game fowls originated in Persia, where they deem it likely that a race of white-legged birds were very early reclaimed. but whose originals, like many wild animals, have long since become extinct. Their sporting history is recorded in Persia, and in the early records of China, although most writers point to Themistacles as the first cocker known to fame, who, some historians

BLACK-BREASTED RED MALAY.

state, received an omen of the success of the army he was
leading, from the crowing of the cocks. But Aoileu, the
author cited, says he saw the cocks fighting. Yet Idomeuses,
long before that time, bore on his shield the effigies of a cock
as a martial bird. History informs us that they were bred
for fighting in the reign of Crœsus, king of Lydia. (A. M.
3426). The ancient Dordanii had representations of cock
fighting on their coins. The fighting cock was one of the
principal gods of the Lyrians, and the learned Hebrew, Dr.
Rabbi David, interpreting the 17th chapter of II. Kings,
verses 30 and 31, says "Nergel" was a cock for war or fighting, or a champion cock, and by the Samaritans worshipped
for a god.—Cocker's Guide.

BLACK-BREASTED RED MALAY.

This breed much resembles the Black-Breasted Red
Game in appearance and is the only variety of Malays that
is recognized by the American Poultry Association, although
white, pyle and black colored varieties are also raised. None
of them have ever been very popular in the United States.
They are ungainly and of a savage disposition; the hens are
only moderate layers and are apt to kill the chicks of others
and sometimes even their own broods. The Malays have heretofore been used principally in crossing with heavier breeds
but they are now being rapidly superseded by their "cousins" the Indian Game. For the excellent illustration of this
breed, on the preceding page, we are indebted to the American
Agriculturist.

BLACK-BREASTED RED GAMES.

This is one of the most beautiful and majestic as well
as best known breed of Games. The accompanying illustration gives a good idea of the general bearing of this bird.
The plumage of the cock, as the name indicates, is red on neck
and back, and black on breast, body and tail. The hen is not
as brilliantly colored; the back is light brown penciled with
dark brown; breast, light salmon, and the body ashy brown.
They are a very hardy fowl and are esteemed for the flavor of
their flesh and eggs, and are perhaps the best adapted of all
Games for the general use of the farmer. Some who have

been induced to keep them will now have no other breed on their place.   They are also favorites with the fancier.

DUCKWING GAMES.

The two varieties which comprise this breed are the Golden and the Silver. They both have the general Game characteristics, and the cocks are alike in the black plumage of breast, body, tail, thighs, and greater part of the wing.

BLACK-BREASTED RED GAMES.

The hackle and saddle of the Golden are straw-colored, and the back, shoulder coverts and wing-bows are golden. The Silver cock has silvery-white plumage in the place of the golden or straw-colored of the first mentioned variety. The hens of both varieties have salmon-colored breasts, black tails, and the rest of the plumage gray, but the Silver hen is of a lighter shade throughout than the Golden.

THE BLACK AND THE WHITE GAMES

are similar to the foregoing except in plumage—that of the Black being wholly black, with metallic lustre, and the White, pure white.

PIT GAMES.

These are often known as "Old English Games," and are

bred in a variety of colors. Their name suggests the use to which they have ordinarily been put, but they are now being more generally bred for their beauty than heretofore. Their eggs and flesh are both of high quality, and though small in size the eggs are produced in abundance. They differ greatly from the ordinary game in shape—the head and legs are comparatively short and the tail well spread and carried high.

JAPANESE PHŒNIX FOWLS

have never been bred any particular color, but are to be seen in nearly the variety of colors that the Games are. The hens have a gamy appearance and the cocks have very long tails, which makes them very attractive on the lawn. They are fair layers, are about the size of the Leghorns, and quite similar to them in habit.

HAMBURGS.

We have already described two varieties of this breed— the Black and the Silver-spangled. There are four others— the Golden-spangled, the Golden-penciled, the Silver-penciled and the White. The only marked difference in the varieties is in the plumage. That of the Golden-spangled differs from that of the Silver-spangled in that the groundwork is reddish or golden bay instead of silvery white, and the tail is greenish black.

The Penciled varieties are much the same in plumage as their Spangled "cousins," except that the black on the hens is in small even bars, and that on the cock is conspicuous for its absence, there being none except a slight marking on the wings and fluff.

The White Hamburg has pure white plumage throughout.

REDCAPS

derive their name from their large, bright red, cap-shaped combs. They have been called the great English layers and are noted for their wonderful egg-producing qualities. They lay a fine large egg and lots of them, some breeders claim more than the Leghorns or any other non-sitting breed, and that they are a better table fowl. They are heavier than the Leghorns and, their flesh being equally as good, they will bring more on the market, after their usefulness as layers is past. This is quite a consideration in the minds of some, but it is impossible to combine excessive egg production and superior table qualities in one fowl. As to whether they are a beautiful bird, tastes differ. Some object to their large combs, and there are breeders making a point of the "neat, medium-sized combs"

REDCAPS.

of their birds. They have a plumage that will stand contamination with dirt, dust and weather the year round, and still look well. They are a shapely, well-formed, compact bird; a fowl for use and commercial purposes. The standard weight of the cock is seven and one-half pounds, that of the hen six and one-half, and when fattened for market they can be made to weigh one-half to two pounds more than the Leghorns. The distinguishing feature of the Redcap, as suggested above, is the rose-comb, which should be at least medium large, full of fancy spikes, stand perfectly straight and firm on the head, with a straight spike behind. The comb of the hen differs from that of the cock only in being smaller. The neck hackle of the cock is a rich dark red, or golden red, striped with bluish black; back, black and red; breast and tail, black; saddle hackle, rich deep red striped with bluish black; wings, nut brown; wing coverts, bluish black; legs, slate color and of good length, free from feathers; ear-lobes and face red. The ground-color of the hen is a rich nut brown, each feather tipped with a bluish black half-moon or crescent-shaped spangle; tail, black; ear-lobes and face, red; neck hackle laced with red. Whether considered ·"handsome" or not they are finding great favor in America; it may be because "handsome is that handsome does."

AMERICAN DOMINIQUES.

From this breed, combined with the Asiatics, sprang the well-known Plymouth Rock.

They are very similar to their progeny, both in good qualities and appearance, excepting they have rose-combs' and are not quite as heavy. They mature quite early and are very hardy, being able to stand our Northern winters much better than most breeds.

SILKIES

are kept for their odd plumage, the feathers being soft, silky and pure white. They have a rose-comb, which is nearly round and of a lumpy appearance. This is generally exposed to full view, as their compact crest falls backward, leaving the face and comb uncovered. The saddle feathers fall on either side of the tail in a silky mass. The shanks and outer toes are covered with silky feathers. This breed has five toes. The hens are good mothers; on this account, and because of the great warmth of their plumage they are used in rearing Pheasants and Bantams.

BLACK SPANISH.

This is the oldest of the non-sitting breeds, having been known for nearly two thousand years. This also makes it one of the earliest breeds that are still in favor. It is a very prolific layer of large, white eggs, but its flesh is not equal to that of many breeds. The birds have a large, smooth face, pure white in color; the contrast between this and their solid black plumage gives them a beautiful appearance to many people. Like most of the Mediterranean breeds their combs are very large and easily frosted; aside from this, they are naturally a hardy fowl, but their hardiness has been somewhat lessened by constant close breeding for a pure

BLACK SPANISH.

white face, showing the foolish mistake some breeders always make—sacrifice a valuable quality to gratify a whim for a fancy (?) point.

ORPINGTONS

are probably Plymouth Rocks colored by Langshan blood, though some claim they are a cross between the Langshan and Minorca. They are fair imitations of the Black Java but inferior to them. The bottoms of the feet are pink instead

of yellow, the color of the Javas.   In England they take the
place of the Black Wyandotte.

ENGLISH "FULL-FEATHERED" COCHINS.

There are "fads" among chicken-fanciers as well as other

FULL-FEATHERED PARTRIDGE COCHIN HEN (ENGLISH TYPE).

people.   One "point" in fancy breeding among English
breeders much sought for is the "full-feathered" type.   Our
illustrations on this and the following page show to what extreme the "point" has been bred.   One is a typical full-
feathered Partridge Cochin, the other a typical Buff Cochin.

FRIZZLED FOWLS

are only "oddities," and may be of any color and with either single or double combs, the one necessary characteristic being that the feathers curve backwards or upwards at the ends, especiallly in the hackle and saddle.

JERSEY BLUES.

This breed is not as well-known as the majority of the other American breeds. In shape and carriage it is quite

FULL-FEATHERED BUFF COCHIN (ENGLISH TYPE).

similar to the Plymouth Rock. The plumage is dark blue except on breast and body, where it is a light shade of blue, laced with dark.

They are not very good layers, and their flesh is rather coarse and stringy.

RUSSIANS.

This is one of the less known breeds. They have a rose-comb and heavy beard. The back is broad and tapering to the tail; the breast, round and full, and the tail is carried erect. The plumage is greenish black.

SULTANS.

The plumage of this breed is pure white. They have

small combs, nearly concealed by their large, compact crests. The beard is full and unites with the crest, thus covering the face. The body is square, deep and carried low. The tail is large and full. The thighs are short and vulture-hocked. The legs are heavily feathered. They have five toes.

BLACK COCHINS.

This variety has the same characteristics as the other

PAIR FRIZZLED FOWLS.

Cochins, but its solid black color gives it the preference with many breeders.

POLISH.

I have already mentioned at length some of the varieties of this highly ornamental breed. The others are very similar, except in color; which is the best is decided by the tastes of the owners. There are two varieties of the Silver, the Bearded and those without beard; their plumage is spangled and laced in the same manner as that of the Golden, but the ground-work is silvery white. The Bearded and Unbearded

White have pure white plumage throughout. The plumage of the Buff-laced is rich buff, each feather being laced with pale buff. They have beards.

RED PYLE GAME.

This breed is noted for their courage and hardiness, yet not lacking in beauty. Color of male, head, hackle and saddle orange, light red or chestnut; back, red or crimson; breast, ground color white; shafts and margin of feathers,

RED PYLE GAME.

chestnut red; wings white and red; tail white; body white. Female, head, brownish red; breast, salmon; rest of plumage mostly white to creamy white.

BROWN RED GAME.

This variety of game is very handsome; color of male, neck, back and saddle lemon, with narrow stripe of black in the middle of the feathers; breast, ground color, black laced with lemon; wing bow, lemon; wing coverts, glossy black; primaries and secondaries, black; body and tail, black. The female, neck lemon with a narrow stripe of black in middle feathers, breast ground color black, evenly laced with lemon; otherwise the plumage is black throughout. The shanks and feet of both cock and hen are dark willow or nearly black. The hen is a very good layer.

RUMPLESS.

This is another oddity and, like the Frizzles, any color or any shape of comb is admissible, so long as there is no sign of a tail.

THE CURASSOW

is a native of South America.   There are more than twelve

THE CURASSOW.

species of them.   The best known is Crested Curassow, which has been domesticated in its native land.   It is of a greenish black color with a white crest.   It much resembles the turkey in size, general characteristics and quality of flesh.   Some claim that the flesh is whiter and of finer flavor. When taken from their home in the northern part of South

America they will thrive in fair-sized flocks in aviaries if kept on dry soil and given plenty of shade and shelter, and given the same care as bestowed upon turkeys. Our illustration is taken from a drawing by Sewell for the Poultry Monthly, of one of a pair on exhibition at Madison Square Garden, New York City, which were much admired and brought out many commendations. There is no doubt about their good points, and it will probably not be long before some live poultry breeder will push them to the front.

WILD TURKEYS,

as found indigenous to North America are the parent stock of all the breeds found in domestication. The male of the common Wild turkey is about three and one-half feet long and five feet in extent of wings, weighing from fifteen to twenty pounds. The naked skin of the neck and head is livid blue and the caruncle purplish red. The general color of plumage is copper bronze with green and metallic reflections, each feather with a velvet black margin; quills brown closely barred with white, tail feathers chestnut narrowly barred with black, and the tip with a very wide subterminal black bar. The female is smaller, usually weighing about nine pounds, and less brilliant in color, without spurs, often without bristles on the breast and with a smaller process above the base of the bill. The legs of both sexes are red or pink.

The gobblers do not get their growth and full plumage till the end of the third year and increase in weight and beauty for several years after that. Gobblers weighing thirty-six and forty pounds have been shot. The feathers lie very close and hard, so that the birds weigh more than their apparent size indicates. The hens do not get their full growth till four or five years old, and may be heavier later. Wild gobblers mate later and the hens lay later than domestic turkeys. The flesh is excellent in flavor and is more juicy and delicate than that of their domesticated descendants. Our illustration is a reduction from a reproduction of Andubon's fine colored plate of free wild gobbler, and is a faithful representation of this noble native American.

WILD TURKEY (MALE).

BRONZE TURKEYS.

The Mammoth Bronze, as it is usually styled, is deemed

BRONZE TURKEY (MALE).

by all to be the largest and hardiest of all turkeys.   The true
Bronze is of a rich, changeable, metallic color, which shines
in the sunlight like gold.   The plumage of hen is not as brill-

iant as the male nor the colors quite as clearly defined though similar throughout. They do not usually attain their full size and weight until from three to four years old. At maturity the hens weigh from fifteen to twenty pounds and the gobblers thirty-five to forty pounds each. The first year, however, they will outweigh any other variety. They are excellent layers, good mothers, hardy, make rapid growth, and are excellent foragers. The young may be raised by hens or turkey hens. The latter are, no doubt, much preferred as the turkey mother is more careful of the tender poults and it has been our experience that they may be entrusted to her care when first hatched, if the weather is warm and favorable. Should the weather be wet about the time they are hatched it is good policy to confine them for a week or more. After this they should be allowed to range with their mother when the grass is dry.

BLACK TURKEYS.

This bird is a favorite on the market on account of its plump body and yellow skin. It is tamer than the other varieties, and fattens on less feed. It is also claimed to be a better sitter. The plumage is lustrous black throughout. Shanks long and stout, of a dark lead or slaty black color. Standard weight: cock, twenty-seven pounds; hen, eighteen.

MAMMOTH WHITE TURKEYS.

This distinct new breed, introduced in 1890, originated as a sport from the Mammoth Bronze, in a similar manner as most white fowls have come as sports from the darker varieties. The breeder has spent a number of years in perfecting them, and now they throw only occasionally a dark poult. They have the general characteristics of the Bronze variety, except that they are even handsomer, mature earlier, and are rather more domestic in their habits. The plumage is pure white throughout, the heads and wattles bright red, and shanks pinkish or flesh color. They almost equal in size the Mammoth Bronze turkeys. They are not, as some might suppose, selected from the White Hollands, but are a distinct breed in every particular, and are certainly a great acquisition as the first and only breed of pure white turkeys that is both hardy and of large size.

WHITE HOLLAND TURKEYS.

White Holland turkeys rank high as table fowls, making a nice appearance when dressed. They have a plump, fine-boned carcass and juicy, well-flavored meat. While not as large as the Bronze turkeys they dress fully as high a percentage of flesh. The standard weight of the cock is twenty-six pounds, that of the hen sixteen. They have a lighter colored skin than other breeds. The skin of the body has a pinkish tinge which is attractive. White Hollands have clear white plumage, though during summer, like other white fowls, they become more or less yellow. The bills, legs and feet are a white, pinkish or flesh color. They are usually favorites on the market on account of their attractive appearance and delicate flesh.

At seven months of age they weigh as much as the Bronze, but at a year old they fall short. They are the best layers and mothers of any of the turkey tribe, and are more domestic in their habits.

NARRAGANSETT TURKEYS.

This breed is nearly as large as the Bronze and matures much more rapidly. The plumage is metallic black, each feather ending in a steel gray band edged with black, which gives it an attractive appearance. The standard weight is thirty-two pounds for the cocks and twenty-two for the hens.

BUFF AND SLATE TURKEYS

are identical with the Black except in color of plumage and shanks. The name in each variety indicates the color of plumage. The shanks of the Buff are bluish-white or flesh-color; those of the Slate are light or dark blue.

ROUEN DUCKS.

This breed is supposed to have originated from the common wild duck. They are greatly admired on account of their changeable colors, with the beautiful markings of the Mallard, the fine close plumage, the rich purple upon the wing of the drake, the delicate pencilings upon his sides, the claret color of his breast, the green and blue reflections of his head and the lustrous green of his back. The color of the female is in general grayish brown with more or less green.

Well fattened, the Rouens are excellent for the table, and hence a profitable market duck. Young Rouens grow very rapidly, and pay well for market, when turned off at an early age; are good layers of large-sized eggs, are very quiet and easy to raise. Pekin and Rouen ducks can be raised

PAIR ROUEN DUCKS.

without either streams or ponds, provided they have plenty of drinking water; and a tub or half barrel sunk in the ground will give them great comfort. The shanks are of a bright orange color. The mature bird weighs more than the Pekin, but does not gain as rapidly while young.

AYLESBURY DUCKS.

This breed is largely raised in England for young market ducks. It attains a large size early and its pure white plumage brings a ready sale for it. When full grown it is larger than the Pekin. The bill should be pale flesh color with no dark spots and the shanks and toes, bright orange. The body is long and well balanced. Where it can have plenty of clear water so as to keep clean, it is pretty.

It is in some respects the most profitable duck for market, as it is better flavored than the Pekin.

CAYUGA DUCKS.

This popular variety originated on Cayuga Lake, from whence its name, and was originally a wild duck frequenting this lake. In the juiciness and richness of its flesh it par-

PAIR CAYUGA DUCKS.

takes of that peculiar game flavor which distinguishes the Canvasback. The plumage of this duck should be a jet glossy black, the feathers of the drake having a lustrous greenish hue in the sunshine, which gives him a peculiarly rich appearance. The Cayugas are very quiet in their habits, and are not disposed to wander from home. They generally commence laying about the first of April, and lay from sixty to seventy eggs before wishing to sit, which they do well, but are careless mothers. They are hardy, and cross well with other ducks. The standard weight of the drake is eight pounds, and that of the duck seven.

CRESTED WHITE DUCKS.

This breed is pure white and has a large, well- balanced crest which makes them very attractive. They are very prolific layers and their eggs unusually fertile. They are a little smaller than most of the other breeds.

MUSCOVY DUCKS.

These in their wild state were originally found in South America. In their native state they are glossy black with

CRESTED WHITE DUCKS.

white wing coverts. They have a naked red face and a large red carbuncle on the top of the bill at the base, and long, crest-like feathers on the head. They have a musky odor (whence their name) which has almost disappeared in the tame duck. There are two varieties in domestication, the Colored—the plumage of which is black with more or less white feathers; and the White—pure white throughout. They are very savage, driving all other fowls away, and are very homely looking. They are the largest of the duck family kept in domestication.

PEKIN DUCKS,

as their name indicates, come from China. They are a large, beautiful bird of erect carriage. The plumage is pure white outside. The inside feathers are slightly cream-colored. The neck is long and gracefully curved, the head long and finely shaped, the eye full and bright.

PEKIN DRAKE.

The legs and beak are a very dark orange, making a fine contrast with the pure white feathers. The mature drake should weigh eight pounds, the mature duck seven. As bred by Mr. Rankin, a pair dressed often weighs twenty pounds at maturity. They lay early and produce more eggs than any other breed. They mature early, are hardy, domestic in their habits, do not wander far and return to their coops at night. They are not mischievous, require less water than other breeds, and their feathers sell at a good price.

To Stop a Stampede.—Pekin ducks are very timid. This sometimes causes trouble when they are closely confined in large numbers. When six or eight weeks old, or even after they are grown, they often get frightened dark nights. Being unable to see, one bird will touch another, it will spring away and touch several more. In an instant the whole flock are in commotion and treading upon each other. There will be a perfect stampede, sometimes kept up the whole night. After such a worrying night many of the birds will be all tired out, and some of them unable to get up. If this disturbance continues—good-bye to all fattening of the birds or any laying. If there is no moon, hanging lanterns about the yards will bring order and quiet.

CALL DUCKS.

There are two varieties of the Call—the White, which resembles the Aylesbury in color, except the bill is bright yellow, and the Gray, which resembles the Rouen. Both are small and kept chiefly for ornamentation in the lakes of parks or the lawns of private residences.

BLACK EAST INDIAN DUCKS.

This is the smallest variety of domesticated ducks. Although the flesh is of fine flavor, they are especially valued for the pleasing appearance of their brilliant greenish black plumage.

EMBDEN GEESE.

These and the Toulouse are the two largest and most profitable breeds of geese, and many consider them as the only ones for practical purposes. The Embden takes its name from a town in Germany. Their pure white plumage makes them especially valuable where feathers are an object. The flesh is very tender and juicy and highly esteemed by epicures, being likened to that of the Canvasback duck. They mature earlier than the Toulouse, but the standard weight is the same, twenty-five pounds for the gander.

TOULOUSE GEESE.

This breed is probably the best known of geese. They take their name from a town on the Garonne r.ver in France. The plumage of the head, neck and back is dark gray, that of the breast and body light gray, shading into white on the belly. They are very hardy and are prolific layers, not being very good sitters. They are more

PAIR TOULOUSE GEESE.

easily kept without a pond of water than the Embden and also with care can be made to weigh more.

AFRICAN GEESE.

The plumage of the African is gray, varying in shade on different parts of the body, They have a large black knob and a fleshy membrane under the throat. The neck is long and well curved but does not have the graceful arch of the swan.

EGYPTIAN GEESE.

This is the smallest of domesticated geese and is also

very prolific. The plumage of head, neck, back and upper part of body is gray and black; the breast is chestnut in center and gray elsewhere, the under part of the body is pale yellow penciled with black, the shoulders are white with a narrow stripe of black.

CHINESE GEESE.

These are, as their name indicates, natives of China. Although they possess much merit, they are kept principally for ornamental purposes. Their swan-like, arched necks give them a graceful appearance in the water. They have a knob at base of bill. They lay large litters of eggs twice or thrice a year. There are two varieties, the Brown and the White—very similar except in color. The plumage of the Brown is dark brown on head and back, and light or grayish brown on neck, breast and body; the knob and bill are dark brown or black; the shanks and toes are dark or dusky orange. The White has a pure white plumage throughout, which, together with their orange-colored knobs, bills and feet, makes them especially attractive. Their feathers, also, bring a high price.

CANADA GEESE.

This breed was formerly known as the Wild goose, being a domesticated variety of the Canadian goose. The plumage of the head, neck and tail is black, that of the back and wings dark gray, breast, light gray, growing darker toward the legs. Under part of body white. The neck is long and slender and the bill and legs black.

CHAPTER II.

CARE AND FEEDING OF POULTRY.

It will not do to say "provide for poultry as nature provides for them," for their conditions and surroundings in domestication are different from what they are in their wild state. Notwithstanding this there is a right and a wrong way to treat poultry if we expect them to do their best and give a money return for their keep. First, they should not be exposed to all sorts of weather with no chance to escape from its bad effects. Shelter from storm and damp and shade from sun should be given. They should have clean quarters, be provided pure water and wholesome food and be afforded opportunity to take care of themselves as nature dictates.

All these requirements may be met, without great money outlay for expensive buildings or elaborate furnishings. The locality and the object for which fowls are kept must largely determine the style of buildings provided and food furnished. In a warm climate the houses need furnish no more than a shelter from rain and wind and a shade from the sun. The food should be determined upon by the result wished and the cost, always seeking a ration that meets the requirements at the least expense. As an aid in determining the latter we give a table on the following page from Wright's Illustrated Book of Poultry, showing the value of various feeding stuffs for feed. Though the table may not be strictly correct as determined by chemical analysis, it gives the relative feeding value of the substances named. In making up rations the elements of "relish" by the fowl and the ability of assimilating must be taken into account as well as "value."

Provided with such shelter as the climate and locality

demand, the next thing needed is to keep it clean. If the
fowls are kept yarded the inclosure must also be kept clean
and healthful by frequently stirring the soil or by a supply
of fresh mold or an absorbent of some kind. Have the floor
of the house dry some way, and if the yard is drained, it will
be all the better; it at least must be free of standing, stag-
nant water. If the shelter is all right and the fowls have a
wide range they will look out for themselves, with much
less work on the part of their owner, who will have only to
see that lime, grit, dust and pure water are where they can get
them. Of course, the nature of the range will determine whether
green food or meat should be provided. If the range be a solid
rock or a sand-hill, even though a quarter section in extent,
the fowls would have to be given some green stuff.

| There are in every 100 parts by weight of | Flesh-forming Material, viz. Gluten, etc. | Warmth-giving and Fattening Materials, viz. | | Bone-making Materials or Mineral Substances. | Husk or Fiber. | Water. |
		Fat or Oil.	Starch.			
Beans and Peas....	25	2	48	2	8	15
Oatmeal............	18	6	68	2	2	9
Middlings..........	18	6	53	5	4	14
Oats	15	6	47	2	20	10
Wheat.............	12	8	70	2	1	12
Buckwheat........	12	6	58	1½	11	11½
Barley............	11	2	60	2	14	11
Indian Corn.......	11	8	65	1	5	10
Hempseed.........	10	21	45	2	14	8
Rice	7	a trace	80	a trace	....	13
Potatoes	6½	....	41	2	....	50⅓
Milk	4½	3	5	¾	....	86¾

Although fowls on a free range will "take care of themselves" at much less cost, in money and labor, to their owner,
it does not prove that they will be more profitable to him
if allowed free range only, for in these days of specialties
the successful poultry-keeper must feed for a special purpose.
The free range fowl will be a healthy one, but the poultryman who makes the most money these days must add to
health some other quality, and his success depends on know-

ing how to feed and to care for his flock so as to keep health, and gain his special point, too. His "point" may be eggs, and eggs alone; it may be early marketable chickens, heavy weight and a fat carcass, or health and vigor of breeding stock and progeny. In each case the feeding and management differ; except in the latter case, there is a choice of breeds for the best results.

One advantage the poultry-keeper has over the keeper of animals for profit is, fowls are omnivorous—eat everything, excelling even swine in this. This fact should teach every one that a constant feeding of one grain or one vegetable or one animal substance will not produce the best results. Yet there are scores and hundreds of farmers who throw out corn, and corn alone, to their fowls from December to April and then declare that "hens eat their heads off every winter." This omnivorous quality of fowls gives the poultry-keeper the opportunity of making up a ration of the foods at his command that will produce the result sought. The far-back parents of our domestic fowls mixed their meat and vegetables as gathered on the range; the successful poultry-keeper of to-day must mix them for his fowls to reach the end sought at the least cost. Variety is the "spice" in a fowl's food. In selecting and mixing the "variety" we are governed more by our surroundings than by choice. The farmer of the West will continue to use all the corn he can consistently with a good ration, while the New Englander near the seaboard will feed all the fish consistent with good results.

Whatever the ration, wholesomeness must be kept in mind. Moldy corn, rotten potatoes and putrid flesh are not wholesome, although fowls will eat them. Perhaps the theory advanced by some that the gizzard removes the objectional features of such feed is correct, but our experience teaches us the flavor, color and quality of eggs are affected by feed. This being the case it does not seem reasonable that all impurities are removed from food by being passed through a hen's gizzard. Experience also teaches that fowls have sickened and died when no cause but improper food could be found. It is safer not to take chances anyway, and it is much pleasanter, to say the least, to eat eggs and poultry not produced from offal, carrion or rotten grain. With the majority of poultry-keepers, grain constitutes the principal part of their feeding ration. at least in money value. Of the

GRAIN

used in this country probably Indian corn outweighs the rest. It is fed whole, cracked, ground, raw or cooked. Referring to the table, on page 77, it will be seen corn contains very little bone-forming material, while it is very rich in fat-forming and warmth-giving substances. Although corn pro. duces eggs with yolks of dark color and rich flavor, it is not recommended for layers unmixed with other grains. For fattening purposes it can not be excelled and should be fed in various forms to keep up the appetite. The "variety" may be increased if some meal is made by grinding the corn and cob together.

Oats are a good nerve food and are not fattening, but their sharpness is an objection to them, as is the amount of waste or useless matter in the husks, especially in poor, light grain. The first objection may be removed by grinding them very fine, but this is difficult to do. Oatmeal is an excellent food but is rather expensive. If oats are to be fed whole or ground husk and all, the heavier they are the cheaper. Forty pound oats contain but little if any more weight of husks than twenty-eight or thirty pound oats. Very light or small oats will often not be eaten unless they are soaked and made larger. This does not add to their nourishment, but compels biddie to get out what little there is in them. If hens that should lay are too fat, a diet of oats will reduce the fatness Ground oats and boiled potatoes make an excellent food for producing fertile eggs and vigorous chickens.

Wheat and its by-products, screenings, bran and middlings, may form a part of an economical ration in many parts of our country, though wheat itself is rather expensive. If screenings are used they should be fed raw so the fowls will not be compelled to eat the dust, poisonous seeds and other foulness contained in them. Moistened bran is apt to produce scours, especially during the winter, and if fed at all should be alternated with whole grain. Though wheat is rich in material for growth, easy of digestion and stimulates egg production, it should be fed less freely than corn, as too much of it produces diarrhea.

In regions where corn can not be successfully grown and barley may, the latter can be used as a very fair substitute; though all that is eaten does not seem to be digested, fowls will thrive on it for a while and it may be used in the make-up of a ration where raised or procured at a reasonable price. There is little value in barley malt; it must be fed fresh. If used too freely it scours.

In this country buckwheat is fed more to make a glossy plumage than as a staple part of the ration. It is very fattening, and in France where largely used it is said to be valuable in whitening the flesh. The yolks of eggs produced from it are pale. Sunflower seeds are also good for giving a glossy plumage and a few fed occasionally whet the appetite.

We have never been able to induce our fowls to eat whole rye, so consider it of no value as "chicken feed." Of course when eaten it has some value, but we should never buy any and try to worry it down our flock for the sake of "variety." Rye feed or bran, mixed with oat or corn meal and moistened, might help to cheapen a ration under some circumstances. It should not be fed alone as it may cake in the crop and produce death.

Millet and Hungarian on account of their small size are very nice grains for young chicks and where raised or when reasonable in price may help make up the variety in the ration of fowls.

In the rice-growing States, that grain is often the cheapest feed that can be procured. This is especially so when broken, or dirty or discolored from wetting. It is claimed to be better than corn meal for young chickens. In India it is much used for fattening poultry. It produces white flesh.

After grains, it is an open question whether green food in the shape of grasses and herbs or animal food in the form of insects comes next in the bill of fare of birds, from which our poultry is derived, in their wild or native state, but we are inclined to the belief that

ANIMAL FOOD

stands first, so place that next in our list of foods for poultry. When insects are abundant and fowls have a large, free range they can generally help themselves to about all they wish, but when confined and during the winter, animal food must be supplied by the keeper if had at all. Even when on the range there are some drawbacks, as certain classes of bugs with hard hooks on their legs are disastrous to young poultry; the hooks fasten to each other and to the crop, causing a fatal distension of that organ. Another drawback is the liability of the fowls eating poisonous bugs and vermin. Again, some breeds are not good enough foragers to get sufficient "meat" to do their best for their owners. And unless the range is very large the best foragers are curtailed in their operations.

If one were to follow the course of nature and furnish poultry the form of animal food they most prefer and than which there is probably nothing better he would give them worms and maggots. These may be bred and fed in their prime; being easily digested they are peculiarly fitted for supplying young chickens with animal food. The breeding places of such food as are offensive to the nostrils should be located at some distance from the house. In summer time expose a quantity of fresh bones to the flies for a day or so, then cover them lightly with fresh mold and in two or three days there will be thousands of worms ready for the fowls. The carcass of any animal either whole or in large pieces hung up out of reach or slightly buried will furnish plenty of maggots. Blood procured from slaughter-houses, exposed to blow flies and then lightly covered with manure, will produce large white worms in abundance. Other ways of obtaining these delicacies for poultry will suggest themselves to him who wishes to minister to the wants of his pets or money-makers. For little chicks there is probably no easier, surer way of providing good, wholesome, palatable live animal food than furnishing them meal or flour worms, which may be done as readily in winter as in summer if the hatchery is kept in a warm place. As a starter, get a few hundred worms from a baker, miller or any one else who has flour stored in quantities. Place the worms in a crock or other earthen vessel with odd bits of woolen cloth rubbed with tallow, crumpled paper and other refuse mixed with musty meal or flour. Over all put some gauze cloth or cotton waste, which must be kept moist. In a couple of months there will be a supply that may be drawn upon daily if the feeding stuffs are renewed from time to time. If chickens are raised in large numbers and it is desired to supply them with flour worms, the size or number of the propagating "house" can be increased. The hatchery for worms may be set going before the incubator or biddie; then there will be no waiting for a supply of insects when the chicks are ready for them.

Without doubt the cheapest and pleasantest way, in the majority of cases in this country, of supplying animal food to poultry is by furnishing them meat in some form. For small flocks, table scraps will generally furnish enough, but where fowls are kept in any number a given amount should be provided. An ounce a day per adult fowl of ordinary size is considered about right. The large Asiatics need a little more. Mr. A. W. Kinney, Yarmouth, Nova Scotia, is an extremist in meat feeding, and, if it were not for getting his hens too fat, would probably feed them meat exclusively. He, in theory, thinks a varied ration is nonsense, but in practice he gives a mixed ration. When his hens are moulting he feeds more than half a pound of meat a day per hen. When his flock consists of hens, through moulting, and young pullets he feeds fifty pounds a day to 200 fowls, and thinks that hardly enough. He uses cows' heads, boils them till the meat all slips off, then runs it through a cutter and mixes with ground grain—one pound of grain to five of the meat. This makes about the right composition. The hens are watched; so is the mixture. If too much grain, they pick out the meat and leave the grain; if too much meat, it will pack down and they will not eat as much as he thinks they ought to have. To still more vary their ration, he cuts up the bone and feeds that for a day or two after feeding the meat for some days. Mr. Kinney says he has "some Light Brahma hens that will take a pound of meat each day for weeks, and not wink at it, but rather look around for several ounces more before roosting time." But they get so fat they will not lay. He says "a hen will grow fat very much quicker on meat than on corn."

Other good feeders claim that meat should be chopped or minced, so each fowl gets its proportionate share. We are a great believer in exercise and we as often feed meat to our flock in large pieces as minced, as it causes the fowls to work more for what they get. The effort in getting it off and the exercise taken in chasing one who has a piece that can be carried aloft in the beak, is beneficial, we think. The same end is reached by hanging a pluck just high enough to make a fowl jump to reach it. Meat furnished by means of cows' heads and sheep's heads, uncooked, gives exercise. Where large flocks of poultry are kept the heads are often boiled and the liquor used in making mush for the fowls. In this case the meat may be chopped and fed separately or as an ingredient of the mush. It goes without saying that fresh meat and not putrid is the proper kind to feed. For little chicks we should mince the meat, whether fed raw or cooked. Do not feed too much to chicks.

The question as to which is the most economical way to provide the meat will present itself to each poultryman, and he must solve it as his circumstances, surroundings and judgment compel. Some may be able to get butchers' waste— plucks of calves, sheep, hogs and cattle; or, perhaps, only the lungs. If not cooked, or fed in large chunks, these may be run through a sausage-chopper. In dairy regions, where the young calves are sold for about what their skins are worth, a good supply can be had cheaply for large flocks. In cities or villages enough may be had for ordinary flocks from the local marketmen—the trimmings made on the block and counter. Before the introduction of machines for pulping green bone we used to procure such at fifty cents per hundred pounds, from a near marketman. After the fowls had cleaned off the meat we gathered the bones and used them as fertilizer. Now, with a green bone cutter, the whole mass —bones, gristle, fat and lean—may be put in the best of shape to have the fowls consume it all. And the "all" is a most excellent food.

Milk is an animal food and one of the best for young chicks. It is not probable that much whole milk will be fed, but skim milk may be made as good, practically, for feeding purposes, by the addition of a little tallow or other fat.

Fish, flesh and fowl are usually considered enough alike to be substituted one for the other, in ordinary households; so, in the ordinary poultry-yard that is situated near the sea, fish is considered a cheap substitute for flesh; but there is one drawback—most people prefer their roast duck or broiled chicken to have a fowl taste, not a fish, and fish do flavor fowl when fed to them in excess. Some claim that the fish taste will not be present if the use of fish as feed is discontinued for five or six weeks before killing the fowl. Eggs are also flavored, which precludes the feeding of much fish to laying hens. If fish are fed raw they should be chopped up and a little salt and pepper added. If boiled, they may be thrown out whole. Never feed stale fish. Where clams are plenty and cheap they may be crushed fine, shells and all, and fed either raw or cooked. After feeding meat in chunks, or whole fish, all refuse should be gathered and consigned to the compost heap.

GREEN FOOD.

If any one doubts that fowls need or relish some vegetable in their diet let him turn a flock from confinement in bare quarters onto a plat of grass; even though there may be corn in abundance in plain view, they will tumble over that and each other in their haste and eagerness to get a nip of the grass. Where abundance of range can be had, the cheapest, best way to furnish green stuff is to let the fowls help themselves to what grass they wish. It is claimed that ten geese require as much pasture as a cow, and two hundred hens will consume or destroy the grass on an acre of ground. At any rate, it is a fact that where poultry is kept in large numbers green stuff must be furnished them, or the buildings will be so far apart as to add so much to the labor of caring for them that the profits will be reduced, or else the poultry will not do their best. In short, the poultry-keeper must supply green vegetable food to his flock, if confined, or if very large, if he means to make money from them. If he is keeping them for pleasure he needs to, that they may be healthy, pretty and spry. How shall this be done? Grass and clover stand first as regards cheapness and ease of supplying in summer, and if cut at the right stage and cured properly they make good winter feed. Green corn and young grain, cabbage, lettuce and fruits, come next for summer food. The vegetables, turnips, beets, potatoes, carrots, come in well for winter feed as do cabbages and apples. Geese will do well upon a ration composed almost wholly of grass —fowls do not do as well if their ration is principally grass or other green food. Some is necessary to good health; too much is not conducive to their best welfare. If fowls have not been accustomed to green food, especially in the winter time, they may be taught to eat it by mixing it with their meal, at first. But if given them in cabbages they will hardly need to be taught. Loose heads, buried in the fall, will be much firmer when taken out in the winter or spring. Hay made from nice grass or clover may be made nearly as palatable as when fresh, by running it through a cutter and then scalding or steaming. The same end may be accomplished with less work with silage. The refuse of the vegetable garden, pea vines and corn husks, may be run through a cutter and put into a silo and used in winter. Whatever is cheapest and handiest is the "what" to use, but be sure and use something. If turnips are used they should be chopped up fine. Kohl-rabi and other tender vegetables will be readily eaten if they are simply cut in two and placed where the fowls can get at them. In winter, when the flock is the least able to supply itself with green food, is when most farmers fail to provide it. A little thoughtfulness and time in the summer and fall will procure an abundant supply for what fowls are kept on the ordinary farm. If no provision has been made, at least a little hay or a few corn-stalks should be run through a cutter and thrown to the fowls two or three times a week.

Young onion tops chopped fine are excellent for young chicks. If to be sold as broilers the onions should be withheld a while before killing, as they impart a flavor to the flesh that those who prefer to do their own seasoning do not relish.   Dandelion leaves are relished by young turkeys, and apples may be chopped and given to geese; hens will take care of apples that are not very hard—harvest apples, for instance—without any preparation, except placing them within their reach.   Young and tender weeds, fresh-cut sods, lawn clippings and ears of corn in the milk or dough all give green food in the summer that is relished by poultry.

COOKED FOOD.

No one disputes the fact that birds of all species in their wild state take their food, be it grain, animal or vegetable, in a raw state—in a wild state for that matter; but our poultry has been bred so far from their natural condition, and so much more is required of them in egg production, weight of carcass or early maturity, that they are called upon to live and work at high pressure, and must have their wants, abnormal though they be, supplied in keeping with the requirements.   One way to do this is to cook part of their food; this alone adds variety if we use but one grain and feed part of it raw and part of it cooked.   Fowls prefer some foods cooked rather than raw; others raw to cooked, and their preference should be consulted.   Care must be used in feeding cooked food to laying or breeding stock, as it is more fattening than raw food.   In cold weather cooked food may be fed warm and is greatly relished.   As cooked food is more easily digested than raw, it is best to feed raw grain at night, as the time till the morning feed is longer than between the other feedings.   Corn is an excellent evening meal and in winter it is well to warm it before feeding.

The simplest way to cook poultry feed is to boil it.   The grains—corn, wheat, buckwheat, rice—may be boiled or steamed.   If boiled they should be kept from the bottom of the vessel by means of a perforated plate of sheet iron.   Mush may be made from any of the grains ground and fed when fresh made or cold.   If fed fresh be sure it is not too hot. Fowls have died from being fed food that was too hot.   Beets, turnips, potatoes, pumpkins, may be boiled, mashed and a fine pudding made by thickening them with meal of any kind, bran or middlings, or a mixture of these.   The pudding will be more civilized if the vegetables are cleaned before being cooked.

Whether cooked or raw, sloppy food is not recommended. Give solid food and drink, either milk or water, but do not compel the fowls to eat a lot of slop to get a little solid. Meals of various kinds, either singly or mixed, or mixed with bran or middlings, may be mixed with water or milk, a little salt added, and baked into cakes.. If water is used, the cake will be much lighter if some cheap baking powder is used. Sour milk and baking soda or saleratus will give much better satisfaction. Cornmeal cakes made this way, then moistened with sweet milk, make an excellent factor in the feed of young poultry. Where milk is fed in large quantities it is much better to scald it. If sour when fed to little chicks they will relish it much better, and will thrive better on it if it is heated enough to separate the whey from the curd, giving them the curd only. Beans and peas are hearty food; the former need to be cooked in order to have the fowls eat them; ground, the meal may be used in mush. Boiled whole, they may be thickened with meal of any kind. Where but a small number of poultry are kept on the farm or in the village, and there are not facilities for cooking, except on the kitchen stove, it is more than likely the cooked food given them will consist principally of mush or scalded meals. In this case the raw vegetables must not be omitted in winter. Where arrangements are such that the cooking may be done without interfering with the household arrangements, it will pay to cook some, even for a small flock.

FEEDING FOR FRAME AND FLESH.

When the chick comes from the shell, the first thing the owner wants of it is to grow—grow in frame and flesh. Then it must be fed for that. It will need no food for the first twenty-four hours; it carries the first day's supply with it from the egg that has nourished it during incubation. Yolk of egg being its infantile food it will be well to continue it a few days. Drop the yolks into boiling water, and when partly cooked mix with an equal quantity of bread crumbs or corn cake made as recommended above Where raised in large numbers bread may be made from cheap flour and with less kneading than for household use. The longer the egg diet is continued the faster and stronger the chicks grow. One egg a day for six or eight will be enough at first; gradually increase the amount till at the end of three weeks two a day are fed to half a dozen chicks. Eggs that have been in the incubator for a few days and proved to be infertile will do good service in this way and should be saved for this purpose.

The feeding should be at regular intervals, and for the first week once in two hours is proper. As they increase in age the number of feedings per day may be lessened till three times daily is reached. When three months old, four or five times a day is better than three. As they grow, and need more than the egg and crumbs, they should be fed the meal worms mentioned under animal food, or liver and other meat may be boiled and chopped very fine and given them. Also, cook coarse corn meal. For variety let them have some millet or hungarian seed, chopped grass, onion tops, boiled potatoes; as they get still larger, wheat, cracked corn and rice. When feathering-out time comes let them have bone meal, or pulped green bone and grissle. The bone meal may be mixed with the regular feed. Let them have a small, clean grit of some kind. And do not give all the skim-milk to the pigs. Fed sweet to the growing chicks it will bring a fair price per 100 pounds. If given all they want of it, they will need no water, except in warm weather. Give it then for their comfort.

FEEDING FOR EGGS.

Before our chicks reach the age at which the pullets will lay, the cockerels, unless reserved for breeders, should have been sold for broilers or spring chickens, or caponized. When the pullets are fed for egg production, two extremes in feeding are to be guarded against—too much and too little. Too much produces fat, and a fat hen will not lay—too little gives insufficient nourishment, and a weakly hen will do little laying. The amount fed is not the only thing to be considered—the quality, as well, has something to do with the laying, and the condition of the hen when through. The draft on her system is not slight, and the amount the hen eats to supply this draft is great. The digestive organs are taxed to their utmost to take care of the extra feed, and the wise feeder will provide the ration that can be easily digested, so as to use all the surplus energy of the hen in producing eggs; not in taking care of whole corn as an entire ration. Give some corn, especially at night, but let the food be varied. Cooked or scalded meal and middlings, cooked and raw vegetables—grass or hay steamed. Provide lime in some shape. Give ground bone; crushed oyster shells; a piece of lime daily in the drinking water; burned bone; refuse from mortar beds. All these will furnish lime, which, with that in the wheat that ought to be fed, will provide enough. If egg shells are fed they must be finely pulverized or the hens may form the egg-eating habit.

Considering egg production for consumption (not hatching) only, the hens may be stimulated somewhat by the use of cayenne pepper or other warming condiments. Some good feeders do not use condiments of any sort, unless salt may be classed as one. It in not a bad rule, if condiments are used, to season the food as you would for your own taste. Now, tastes differ, but the longer one uses condiments, the stronger or thicker he wishes them—follow the same plan with the fowls. Always season with a little salt whether you believe in condiments or not.

Rock salt, or salt that contains large crystals, should not be exposed so fowls can help themselves, as they would be apt to help themselves to it for grit, and it would not take long for an injurious, if not fatal, amount to be swallowed. Experiment has shown that a quarter of a pound of salt may be fed to 100 hens each day without injurious effects, after they have been fed a smaller amount for some days previous. It is probable that an ounce a day for 100 mature fowls is about right for health and best results. Laying hens, especially those confined, should not be fed too much tallow. Experiment has proved that hens with oil meal instead of tallow in their rations laid a few more eggs; tallow is also deficient in nitrogen and the moulting season is delayed and prolonged.

A highly nitrogenous ration helps to early, speedy moulting. Give warm, clean quarters, variety in feed, pure water, grit and lime to young hens or pullets of a laying breed, and good eggs in abundance ought to be had.

FEEDING FOR FAT.

Food is not all that is needed to make fat fowls. The fowl must be well and well-fed up to time of being fed for fattening. To get the best return for food consumed the fowl should have obtained its growth before fattening begins. In fact, before this, one can hardly be said to be fattening the fowls, though in a well fowl all food that is digested goes toward frame or foundation for building fat upon. And if a fowl has not been well fed from the shell up, it never can be well fattened. The fat will be poor or the expense so great for getting on good fat that the fattener will be poor—in pocket from the transaction.

Well-fed chicks of the large breeds will be ready for the fattening pen when four or five months old. Some cross-bred fowls will be ready for the pen earlier than this if properly fed. We say ready for the pen, because quiet and inactivity are conducive to the best results in laying on fat, though the flavor of the flesh of fowls confined is not equal to that of those allowed to run at large. Put ten or a dozen in each pen, hens by themselves and cocks by themselves, putting together those in each case that have been accustomed to being together. Have the coops warm and without perches; be sure they are clean, dry and kept free from vermin. They should be darkened at least part of the time. If in a darkened building that is enough; if not, a blanket may be thrown over the coops for two or three hours after each feeding. The ration must contain but few vegetables, and little green food or milk. The meat food should be largely fat, but this should not be fed in excess, as it may lessen the fine flavor of the flesh. The experience of Mr. Kinney contradicts the anti-meat theory, as he is able to lay on fat with lean meat alone. But this would not be an economical ration in most localities in our country where corn is recognized as the cheapest and best food for fattening purposes. This being so, the only thing to be guarded against is too constant feeding of a uniform ration. Vary the ration often enough to prolong the relish for feed and the power of digestion till the fowl has reached the highest possible point of profitable fatness. This may be done by changing from whole corn to meal, fed either raw or cooked. Oats and buckwheat may be fed as a change. But corn meal is the staple feed; the others, with cooked potatoes or other vegetables are used only to give variety. Let all be done that can be to keep up the appetite. A few hours' fast after first being put in their quarters will give a good send-off for their first meal, which should not be more than they will eat up clean. Let the morning meal be given as early as possible, then feed at intervals of four hours through the day,giving the last one late at night. A little salt and pepper may be used, also charcoal. Milk may be given as a drink or used in mixing the different meals. If care has been used in the feeding so that fat has been laid on with a good degree of success, the fowl may as well be killed and marketed as soon as it gets off its feed or when it begins to fall off. Unless turkeys are "crammed" they should be fattened while allowed free range. Shut up those that are to be held for breeding purposes or for an aftercrop and feed the fattening ones at intervals of four hours through the day all they will eat up clean. Leave no food by them.

HOW, WHEN AND WHAT TO FEED.

Birds in their wild state get their food slowly and a little at a time. It is well that fowls get their food the same way. It is not a good plan to have food before them all the while; so, excepting soft food, which may be given in troughs, it is best to scatter their grain rations among straw, leaves or in light soil and place their animal and green food ration where they can pick at them and gather what they want at leisure and with exercise. The V-shaped trough made of six-inch fencing is all the utensil we consider necessary to feed from; if of dressed lumber it can be more readily kept clean.

As to when to feed, breeders differ. Some claim that adult fowls should be fed three times daily; others hold that twice a day is enough. Both classes admit that the last feed

should be just before roosting time.   Young chickens ought
to be fed at intervals of two hours at first.   The period between feedings may be lengthened till they are three months
old, when three times are enough and if twice is enough for
adults it is about time to break the "chicks" to that course
too.   If fed three times there is more danger of overfeeding
than when fed twice, especially if on the range; and overfeeding is really more disastrous than underfeeding, as there
is usually a chance to more or less supplement the short feed.
To feed just the right amount is more important than the
number of times at which it should be given.

The "what" to feed has already been discussed and
answered.   Feed a variety—grain, green food and animal
food.   Feed some of each every day.   Because this is accomplished where the small flock is kept and given the table
scraps accounts for so many "best egg records" being made
by a small number of hens.   Multiplied by hundreds, in
theory the results should be increased just as many fold.   In
most cases this does not prove true, because the same variety
is not maintained, though the same care otherwise is given.
As has already been stated, there is one other element besides "variety" entering the answer to what shall be fed—
and that is cost of rations.   Feed variety at the least outlay,
quality considered.   These two elements open up a wide
range for the ingenuity, thought and judgment of the
feeder.

CHAPTER III.

DRESSING AND SHIPPING POULTRY.

Poultry should be kept without food or water twenty-four hours before killing for market; full crops injure the appearance and are liable to sour, and when this occurs, correspondingly lower prices must be accepted than obtainable for choice stock. Never kill poultry by wringing the neck. The demands of various markets vary a little in the manner of dressing poultry, and in preparing it for market, the custom of the market to which one is to ship should be followed.

CHICKENS FOR CHICAGO.

Kill by bleeding in the mouth or opening the veins of the neck; hang by the feet until properly bled. This is best done as recommended in directions for dressing capons. Leave head and feet on; do not remove intestines nor crop. Scalded chickens sell best to home trade, and dry picked best to shippers, so that either manner of dressing will do if properly done. For scalding chickens the water should be as near the boiling point as possible, without boiling; pick the legs dry before scalding; hold the fowl by the head and legs and immerse and lift up and down three times (if the head is immersed it turns the color of the comb and gives the eyes a shrunken appearance, which leads buyers to think the fowl has been sick); the feathers and pin-feathers should then be removed immediately, very cleanly and without breaking the skin; then "plump" by dipping ten seconds in water, nearly or quite boiling hot, then immediately into cold water; hang in a cool place until the animal heat is entirely out of the body. To dry pick chickens properly, the work should be done while the chickens are bleeding; do not wait and let the bodies get cold. Dry picking is much more easily done while the bodies are warm. Be careful and do not break or tear the skin.

TURKEYS FOR CHICAGO.

Observe the same directions as are given for preparing chickens, but always dry pick. Dry picked turkeys always sell best and command better prices than scalded lots, as the appearance is brighter and more attractive. Endeavor to market all old and heavy gobblers before January 1, as after the holidays the demand is for small fat hen turkeys only, old toms being sold at a discount to canners.

DUCKS AND GEESE FOR CHICAGO

should be scalded in water of the temperature as for other kinds of poultry, but it requires more time for the water to penetrate and loosen the feathers. Some parties advise after scalding, to wrap them in a blanket for the purpose of steaming, but they must not be left in this condition long enough to cook the flesh. Two or three minutes is the time recommended. Do not undertake to dry pick geese and ducks just before killing for the sake of saving the feathers, as it causes the skin to be very much inflamed, and is a great injury to the sale. Do not pick the feathers off the head; leave the feathers on for two or three inches on the neck. Do not singe the bodies to remove down or hair, as the heat from the flames will give them an oily and unsightly appearance. After they are picked clean they should be held in scalding water about ten seconds for the purpose of plumping, and then rinsed off in clean cold water. Fat, heavy stock is always preferred.

CAPONS FOR CHICAGO.

Capons are dressed in a manner peculiar to themselves. When the dressing-place is selected, drive two spikes about a foot apart in a beam overhead. Then make two loops of strong string, each long enough to hold one leg of the capon, and when hung from the nails above let the bird hang low enough to make picking handy. Have a weight of two or three pounds with a hook attached. When the bird is killed fasten the hook into his lower jaw to hold him steady while picking.

When you are ready to kill your capon, catch him, and if his feet are soiled wash them, then suspend him by the two legs from the nooses. Take hold of his head, and with a small sharp knife cut the vein at back of throat, through the mouth. Never do this from the outside. As soon as you cut the vein, run point of knife through the roof of mouth into the brain. As soon as the knife enters the brain, the bird loses all sense of feeling. Begin plucking at once.

When the plucking begins is when the "dress" of the capon begins to show itself. The feathers are left on the wing up to the second joint. The head and hackle feathers, also those on the legs half way up the drum-sticks, with all the tail-feathers, including those a little way up the back, and the long ones on the hips close to the tail, are left on.

This manner of dressing, with the peculiar looking head, are the distinguishing features of capons, which enable them to be readily identified among a host of other fowls. The plumage being heavier than that of cockerels, and the smallness of the comb and wattles, which stopped growing as soon as caponizing was performed, prevents the palming off of any cockerels as capons by evil-disposed persons, even should the style of dressing be copied.

Care should be taken that the capon is not torn in plucking; when this is completed wash the head and mouth well with cold water, being careful to remove all blood. When cool they are ready to pack for shipping—for some markets they must be drawn before packing.

When capons are to be drawn, have a table the right height to work at handily, and on it have a frame made like a small box with the cover and two ends knocked out. Take the weight off, and put the bird, back down, in the frame. Cut carefully around the vent and pull out the intestines. When the end of the intestine is reached, put your fingers up in the fowl and break it off, leaving everything else in. There will be considerable fat around the opening made; this should be slightly turned outward; it will soon cool, become hard, and add to the rich appearance of the bird. Now hang the bird in a clean cool place till thoroughly cold.

When cold they are ready to pack. Have new boxes of a

proper size for the number you wish to ship, but do not have the packages overly large; line them with clean, plain white or manilla paper, and pack the birds in solid, back up, but do not bruise them. Put paper over them, nail on the cover, and mark as directed under

PACKING AND SHIPPING.

Before packing and shipping, poultry should be thoroughly dry and cold, but not frozen; the animal heat should be entirely out of the body; pack snugly in boxes or barrels, but use great care to avoid bruising the flesh or breaking any bones; boxes holding 100 to 200 pounds are preferable; straighten out the body and legs, so that they will not arrive very much bent and twisted out of shape; fill the packages as full as possible to prevent moving about on the way; barrels answer better for chickens and ducks than for turkeys or geese; weigh the package before packing; when convenient, avoid putting more than one kind in a package; if more than one kind in a package, mark kind and weight of each description on the package; if but one kind in the package mark in plain figures on the cover the number and kind of birds within, the total weight of package and net weight; mark shipping directions, your name and address, plainly on the cover. They are then ready for transporting. Mark name and address of firm to which they are to go plainly on cover, and send full advice and invoice by first mail after the goods are shipped.

THANKSGIVING DAY POULTRY.

For a week previous to Thanksgiving Day a large trade is expected in turkeys in Chicago, and a few points in regard to this special week may not be amiss right here. In the first place take special care and ship only choice, large, well-fatted turkeys, weighing not less than ten pounds, and from that upwards. Small, light weights sell best around Christmas and after New Years. What merchants particularly request is not to ship scalawag stock, as it is not wanted, and only very low prices will move it, and there is liable to be more or less dissatisfaction. Do not kill poor, thin turkeys. It will pay to hold such stock back until it acquires more flesh. Turkeys, too, should be dry picked; they have a better appearance and sell better for shipping purposes than scalded.

In the second place, do not wait till a day or two before Thanksgiving Day before sending in the stock, expecting that the time to sell turkeys to the best advantage. Years of experience have taught merchants here that a week, even ten days, before Thanksgiving Day is very often the best time to have turkeys on the market, for the reason that shippers receive orders from all quarters and this stock must be bought and shipped in time to reach destination at least a day or two before Thanksgiving Day. This competition among shippers generally results in developing a strong market and high prices, while on Thanksgiving Day there is only the home demand to depend on, and the supply then invariably exceeds the demand. Therefore, it is advisable to ship right along, and not have poultry intended for Thanksgiving Day trade come later than the Monday previous. Of course, should the weather be mild, exceptions should be taken to the above instructions, but if the weather is cold and favorable, ship right along. Remember, however, keep small, thin turkeys at home. Usually but few chickens are wanted, and better let some one else take the chances on shipping them, as they are usually a drug on Thanksgiving Day. Ducks sell only moderately, and geese very slow.

POULTRY FOR BOSTON.

Boston is a good market for poultry that has been properly prepared. If shippers desire to realize full market prices for their consignments, they must see to it that their poultry is of the quality and in the condition to suit the best class of trade.

The style of dressing for the Boston market differs from the Chicago style in that all poultry should be killed by bleeding in the neck, and picked while the body is warm; in no case should it be scalded; wet-picked poultry is not wanted in the Boston market, and will not sell for what it is really worth.

As soon as the poultry is picked, take off the head at the

throat, strip the blood out of the neck, peel back the skin a little, remove a portion of the neck bone, then just before packing, except in warm weather, draw the skin over the end and tie and trim neatly. Draw the intestines, making the incision as small as possible, and leave the gizzard, heart and liver in. Pull out the wing and tail feathers clean.

Undrawn poultry can be sold to a limited extent when there is no other to be had, provided there is no food in the crop or entrails, but as a rule it has to go at very low prices.

GAME FOR BOSTON.

Grouse and quail should be carefully wrapped in paper and packed in small boxes or barrels, with the heads down. Never in any case should the entrails be removed. Mark the number of grouse or dozen of quail on each package.

PACKING FOR BOSTON.

After the poultry is entirely cold sort it carefully, and have the No. 1 stock of uniform quality. Pack the No. 2 stock in separate packages. If you have any old tom turkeys put them in a separate package or with the No. 2 stock. Line the boxes with clean paper; never use straw in packing, and never wrap the birds in paper. Pack as closely as possible backs upward, legs out straight, and see that the boxes are so full that when the covers are nailed on, there can be no possibility of the contents shifting about. Boxes are the best packages, and should contain from 100 to 200 pounds. The directions for marking and shipping given under Packing and Shipping, page 96, are applicable to the Boston market.

THE LAW.

The following law regulating the sale of dressed poultry in Massachusetts explains the why for some of the requirements of the Boston market, and indicates the trouble or expense a shipper causes his commission man by sending poultry that is not dressed "according to law:"

SEC. 1. No poultry, except it be alive, shall be sold or exposed for sale until it has been properly dressed, by the removal of the crop and entrails when containing food.

SEC. 2. Whoever knowingly sells or exposes for sale poultry

contrary to the provisions of Sec. 1 of this act, shall be punished by a fine of not less than five nor more than fifty dollars for each offense. The boards of health in the several cities and towns shall cause the provisions of this act to be enforced in their respective cities and towns.

POULTRY FOR PHILADELPHIA

should be fat. Do not send any poor birds there, no matter how high poultry is selling. Kill and dress as for the Chicago market, except both turkeys and chickens should always be dry picked. To get the animal heat out hang up, heads down, for twelve or fifteen hours, or put in ice water long enough to make thoroughly cold, and then hang up till perfectly dry. Ducks and geese should be full feathered before killing. Turkeys and chickens that are very fat and handsome will command a little higher price.

Best markets for poultry are Thanksgiving, Christmas and New Years. Care should be taken that poultry shipped for these special days arrive in sufficient time before the date to meet the best sale. Poultry that arrives too late generally meets a poor market. Turkeys will sell on either of these occasions, especially fat hens and young toms. After the holidays, small turkeys have the preference. Geese sell at Christmas, and fancy ducks sell well at any time. Ship capons in cold weather only.

We again say, use great care in selecting, dressing and packing. A handsome appearance is in all cases worth one to two cents per pound in selling. There is nearly always a large supply of poultry during Thanksgiving week, and buyers having knowledge of the fact naturally pick out the best stock to the neglect of poorly dressed, which must then be sacrificed for what it will bring, and the commission men can assure you it is not pleasant to have shipments of that kind.

PACKING FOR PHILADELPHIA.

The general directions for packing, marking and shipping apply to poultry designed for Philadelphia. We mention the following as specially applicable to that market:

For turkeys and geese, boxes are best, although large barrels may be used by experienced packers. Use clean packages, and paper or clean, dry straw free from dust. Place a layer of clean straw at the bottom, then alternate layers of poultry and brown paper or good, clean straw; stow each bird snugly, back upward, legs out straight, and fill the package so full that the cover will draw down tight and snug upon the contents to prevent shifting and shucking while in transit. Barrels are better for chickens and ducks, and these may be assorted and tied up in pairs.

GAME FOR PHILADELPHIA.

Prairie chickens and quails should be packed in boxes with holes in them; the former ten to twenty pairs in each, and the latter ten to twelve dozen. Neither should be drawn; the feathers should be smoothed down; pack with breasts up. Rabbits and hares should have the entrails removed. Genuine wild turkeys bring good prices at the holidays, and should have the feathers left on. Whole deer should have the liver and lungs removed, or allowance must be made in the weight for them. Saddles should be sewed in clean cloth in order to keep them in good order.

POULTRY FOR NEW YORK.

Poultry of any description for this market may be scalded or dry picked, but it sells quicker dry picked. It must be undrawn and heads and feet on.

POULTRY FOR MINNEAPOLIS.

Poultry for Minneapolis, Minn., may be dressed, packed and shipped as for the Chicago market, except that chickens and fowls should have their heads cut off after plucking, leaving the necks of the birds as long as possible. If fowls are well fatted, the trade does not object to taking them if scalded, at same value as dry picked. Turkeys should be dry picked. No objections to tip of wing feathers being left on the birds. The trade has been educated gradually to take undrawn stock, and it is now preferable to drawn stock.

POULTRY FOR ST. LOUIS, MO.,

must be drawn; heads and feet off. Capons are not yet received at this market. Game should also be drawn in warm weather, undrawn in cold. Rabbits drawn all seasons.

PACKING BROILERS FOR MARKET.

Before packing broilers in barrels for shipment to market precaution to remove all animal heat must be taken. It may be done by placing them in ice water after they are picked, allowing them to remain in it ten or twelve hours, then removing and hanging up by the feet in a cool place to drain. After this wipe them dry with a clean towel, and put a layer of broilers in the barrel, then a layer of ice, broken in pieces the size of a turkey's egg, covering the broilers well with the ice, followed by alternate layers of broilers and ice until the barrel is full, which should be covered with clean muslin and a thickness of bagging. Do not begin to pack until you are nearly ready to ship, and have all arrangements made in advance with your merchant. Ship by express and avoid all delays. Never ship so as to reach the market on Saturday, as a portion of the stock may have to remain over Sunday before being sold, which adds to the expense of handling. Besides, there is often a shrinkage in value on account of being held over.

SHIPPING LIVE POULTRY.

There are a few general points in regard to shipping live poultry that are applicable to all markets:

1. Shippers should see that the coops are in good condition before using, so that they are not liable to come apart in transit, as they are roughly handled sometimes.

2. The coops should also be high enough to allow whatever kind of poultry is shipped, room enough to stand up. Low coops should not be used, as it is not only cruel, but a great deal of poultry is lost every year by suffocation. Coops should not be overcrowded.

3. In shipping hens and roosters they should be kept separate. Nothing depreciates the value of a fine coop of hens as much as to have a number of old cocks among them. Shippers often wonder why they do not get the highest market price for their stock; in most cases this is the reason. Good stock always commands a quick sale at best prices.

4. Poultry should be shipped so as to arrive on the market from Tuesday to Friday. Receipts generally increase

toward the end of the week, and there is enough carried-over stock on hand Saturday to supply the demand. Merchants, rather than carry stock over Sunday, will sell at a sacrifice, as the stock, when in coops, loses considerable in weight by shrinkage, and does not appear fresh and bright. Besides, Monday is usually a poor day to sell poultry.

SHIPPING BREEDING POULTRY.

Some poultry breeders ship the fowls which they sell, in any box that comes handy. Sometimes these boxes are just right in size, sometimes they are too small; usually they are altogether too large, and the unlucky buyer pays express on a lot of unnecessary lumber. Do not use old boxes. Have regular shipping coops. These coops should be strong, light, neat and attractive, and large enough for the fowl or fowls, as the case may be, to sit or stand side by side comfortably, but no larger. For the bottoms and the ends of the coops, use half-inch boards; on the sides, at the bottom, nail a strip of board about six inches wide, and another six-inch strip across the top for marking. Make the rest of top and sides of lath. Cover the bottom with cut straw, sawdust, or other light litter. Tack a tin cup inside for water, and fix a place for feed. For long distances make a "hopper" so that the feed will work out at the bottom as fast as consumed by the fowls. Corn is the best food. Fasten to the coop a request to train hands to supply water. If such coops were used there would be less complaint about the express charges.

CHAPTER IV.

DISEASES OF POULTRY.

As nearly all diseases of poultry are caused by cold, wet, want of cleanliness or improper feeding, it is much easier and cheaper to guard against disease than to cure it after it once gets into a flock. It is often hard to locate the trouble or determine the ailment. A hen is moping about with feathers rough, comb dark, appetite poor or gone. The hen is sick, but all of the symptoms may come from any one of a number of causes, and one is at a loss to know what the real cause is. In the case of common fowls we believe the cheapest remedy—we know it is the most effectual—is to use the hatchet at once. If the quarters are not dry and clean, and the proper food and drink have not been given, these should at once be remedied and the "run" of the disease cut short· If the requirements for health have been met we may be pretty sure that the whole flock is better off with the sick bird out of the way. If pure air is furnished poultry, with no draft, wholesome food supplied, and the quarters are clean and dry, the cases of death from sickness will be very rare.

In case of accident or where valuable birds are ill it may be advisable to employ remedies. In case the disease is the result of neglect it is well to use the means that will the soonest get the bad effects of the neglect out of our flocks, so we name some of the diseases poultry are subject to, with preventives and remedies.

Have a pen or room, with a small run, apart from the flock, in which sick fowls may be placed while treated. Always disinfect after the death or removal of a fowl that has had a contagious disease. Have, as hospital stores, castor oil, coal oil, sweet oil, carbonate of soda, carbolic acid, charcoal, Douglass Mixture, pulverized chlorate of potash, roup pills, sulphur, and tonic powders.

ABORTION.

Sometimes hens when driven violently about drop suddenly either a perfect or a soft egg and afterwards mope about as if very ill. In such cases, which are not to be confounded with merely laying soft-shelled eggs, the hen should be put into the pen; have it darkened. A little carbonate of soda may be put in the drinking water; the food should be soft and given sparingly. This treatment should continue a few days unless the patient recovers sooner. Abortion has in rare cases been caused by ergot on the grass runs. In such cases the cause must be removed.

BLACK-ROT.

The usual causes are improper food and filth. The first symptom is usually blackening of the comb. This is followed by swelling in the legs and feet accompanied with gradual wasting away. Treatment to be of any use must be given in the early stages. First give a dose of calomel or castor oil. Follow this with Douglass Mixture or any other simple tonic treatment with warm and nourishing diet. Take better care of those that are left.

BRONCHITIS.

Frequent coughing distinguishes this disease from a simple cold in the head. Remove the fowl to the pen, which should be moderately warm. Add just enough sulphuric and nitric acid and white sugar to the water to make the whole slightly sweet and acid, or give roup pills. A little cayenne or ginger may be added to the feed with advantage.

BUMBLE-FOOT

is a corn or abscess on the bottom of the foot. If it seems constitutional with some breeds, notice if it is not the large ones and if the afflicted birds have not been compelled to roost upon narrow, ill-shaped perches or to jump from high ones to a hard floor. These are most often the causes. Occasionally a case is so far advanced that a cure is impossible but generally if taken in good time a daily application of lunar caustic or painting with the tincture of iodine will effect a cure. In cases where the tumor is full of pus or in the form of an abscess it must be lanced and the matter pressed out, the part fomented with warm water and after a day or two the caustic applied. If the tumor is hard make the incision the shape of a cross-and squeeze out the matter. During treatment great gain will be made if the bird is compelled to "roost" on the floor covered with litter of straw or leaves. To prevent bumble-foot provide flat perches close to floor or with an approach formed by a ladder.

CATARRH OR COLD IN THE HEAD.

A common cold shows itself by more or less discharge from the eyes and nostrils. It is not dangerous, but if neglected may run into roup. Remove to a warm place. Give three drops mother tincture aconite in half a pint of drinking water. Feed moderately on soft food, mixed warm and seasoned with No. 1 mixture, given on page 123. If no better, or if worse in a few days, treat as in roup.

CONSUMPTION.

Damp or cold, want of light, or constitutional debility are the causes. When the cough of bronchitis becomes chronic, with wasting away and loss of strength, this disease may be suspected. We know of no cure. When danger is suspected, a tonic may be given as advised for debility, with hope that the disease may be warded off, yet we should dislike very much to breed from a suspected bird, and should use the hatchet.

CROP, SOFT OR SWELLED.

In these cases the crop is distended, but the contents are soft or fluid. The cause is supposed to be excessive drinking after prolonged thirst, which causes the inner coats of the crop to lose their "tone." They are unable to contract properly on the food and the crop remains distended, even with air. Where taken in hand soon after the attack it can usually be mastered. Put the bird by itself and feed three times daily with a small quantity of soft food thoroughly cooked. Allow it to drink moderately after each meal only, water made slightly acid with nitric acid. Do not leave the fountain in the pen. Season the food with some such mixture as No. 4, page 123. Chopped onions or garlic will be the best green food; in fact, they of themselves have a remedial effect.

DIARRHEA

may be caused by any sudden change in the diet, or even of the weather. If the looseness is observed early it can usually be checked at once by giving a meal or two of boiled rice, sprinkled over with finely powdered chalk. If this does not effect a cure, six drops of camphor may be given three times daily. Restrict the diet to boiled rice and a little cut grass, daily. In very severe cases give half a grain of opium night and morning in a soft pill. When the patients are chicks keep them warm and give each chick a grain of pulverized ginger in the food once a day, and put a teaspoonful of alum in each quart of drinking water.

DYSENTERY

is diarrhea developed to the stage that the evacuations are mingled with blood. It is seldom cured. Good results have been obtained by giving twice a day five drops of laudanum.

DISEASES OF DUCKS.

Ducks are not as subject to disease as hens, and are entirely free from lice and body parasites. Yet cleanliness, plenty of pure air and water increase egg production, promote growth and improve the quality of ducklings. The building should be dry, clean and sweet but not too warm nor subject to too great range in temperature. If the food is healthy and nutritious the greater the variety the better.

Death from Insects.—Many ducklings when allowed free range during warm weather die from devouring injurious insects. Bees, wasps, hornets and bugs of all descriptions are eagerly swallowed alive, and the ducklings often pay the penalty with their lives. To prevent this loss always confine ducklings, even when designed for breeding purposes, till they are six weeks old, after which they may be allowed to range.

Diarrhea.—Young ducklings sometimes have diarrhea. It is caused more by overheating brooders and the exhausted condition of the mother bird than from improper food. It is usually prevalent during warm weather. Do not overfeed or overheat ducklings. The remedy: Feed bread or cracker crumbs, moistened with boiled milk, into which a little powdered chalk has been dusted.

Enlarged or Abnormal Liver is the most dangerous disease to which young ducks are subject. It is seldom prevalent except during warm weather, and usually in ducklings from two to six weeks old. The livers of the little fellows often enlarge so as to force up their backs—a deformity which clings to them through life. It is caused by a complete stagnation of the digestive organs, and often appears after a heavy rain or long wet spell, which makes the yards wet, sloppy and offensive. The ducklings will, while in constant contact with this mud, absorb more or less of it, clogging the digestive organs, and deranging their appetites. Remedy: Remove the ducklings to a dry, shady place, feed sparingly and give a little Douglass Mixture in the drinking water.

Sore Eyes.—Ducklings are sometimes troubled with sore eyes. The adjacent parts become inflamed, the head slightly swelled. The cause is filthy quarters or feeding sloppy food. The feathers around the eyes become filled with the food and the dust adheres to them. This naturally inflames the eyes. Remedy: Wash thoroughly clean with warm water and bathe the eyes with a little sweet oil.

Chills.—Before fully feathered, ducklings are liable to chills if allowed free access to streams or ponds, or to waddle about in cold, wet grass. Treat chilled ducklings as you would chilled chicks.

An Unnamed Disease.—From the Poultry Keeper we take the following: "My ducks act as though they were broken down in the back. They can use their feet and legs, but cannot walk. Put them on their feet and they will fall over. Will eat well for three or four days, then their appetite fails, and they finally stop eating altogether, and seem to dry up. Have nothing left but bones and feathers. They live for ten or twelve days. They give a coarse, croaking noise when you go near them, something like a young bullfrog. Their digestive organs seem to be good, as their droppings are the same as the well ones, as far as I can see."

The editor answers: "There are three probable causes. Damp sleeping quarters, ducks very fat, and injury by the drake's attentions. There are many sources for difficulty, however, such as the eating of some poisonous weed, or injurious substance, the depredations of parasites, etc. The principal cause of sickness among ducks, however, is damp sleeping places."

ERUPTIONS OR WHITE COMB

often result when green food is not supplied in abundance. There will be a whitish scurf or dandruff which if not checked extends down the neck, causing a loss of feathers as far as it goes. Fowls kept in small or dirty yards may acquire a scurfy skin of the same nature. The treatment is similar in both cases. Green food must be supplied and cleanliness attended to. Dress the affected parts with tar and sulphur ointment, or with an ointment made by mixing a quarter ounce of turmeric powder with an ounce of cocoanut oil. This last is a specific in true "white comb." As internal treatment give a dose of castor oil to commence with, follow with a teaspoonful of powdered sulphur daily in the food for ten days. Should the sulphur cause irritation before a cure is effected, as will be indicated by the fowl constantly scratching its head, the parts may be dressed for a few days with sweet oil to which a few drops of carbolic acid have been added.

FROST-BITE.

The best treatment for frost-bitten combs or wattles is to thaw out with applications of snow or cold water. Afterwards apply glycerine or sweet oil daily. Prevention is better, and in most cases may be accomplished by oiling the combs and wattles with a sponge or soft flannel every morning. This not only protects the tissues, but prevents water adhering to and freezing upon the wattles when the fowls drink. No one will suppose that a little oil is to take the place of comfortable quarters, but in case of a sudden severe spell the oiling precaution will save pain and disfigurement to an innocent bird. Another thing, it is claimed a hen will not lay during the time a frost-bite is healing. A fowl with frozen feet should be killed. If done before the feet thaw out and the fowl becomes feverish, it will be all right for table use.

GAPES

afflict chicks, and are caused by small, thread-like worms that get into the windpipe and choke the chicks. Gapes are seldom found where the chicks are on high, dry ground, have good food, pure water, and strict cleanliness about the coops and runs. Place their feed upon clean boards. The addition of Douglass Mixture to their drink is an added preventive.

When once a brood is infected there are several ways of getting rid of the pests. One is to take each chick, and with a horse hair doubled so as to form a loop, swab out its throat. Run the loop down the throat and give it a twist before removing it. Continue this till all the worms are removed.

Some consider the fumes of burning carbolic acid the best remedy. Fix a box or coop so the chicks can be shut in the upper half, then put a few drops of the acid on a red-hot shovel and place it in the lower part, under the birds. Keep the chicks in the smoke till they are nearly suffocated, but keep a lookout that you do not quite choke the life out of them. There are some who use sulphur in the same way with good success.

Another good remedy is to place the chicks in a close box, cover it over with cheese-cloth, and over this put air-slaked lime. Shake a little, so the fine lime will sift down among the chicks, taking care not to overdo the matter and smother them.

If taken in hand as soon as the gasping is noticed, a small bit of camphor gum or two drops of turpentine mixed in soft food and given as a pill will generally effect a cure. If these do not, increase the dose.

Other remedies are: putting good-sized lumps of camphor in the drinking water; mixing garlic or onions freely in their food, or by mixing powdered asafetida and powdered gentian with it.

Chicks that die from the gapes should be burned or

buried deep with plenty of quicklime. Move the coops to fresh ground, and spade the old places up and scatter quicklime freely about the whole premises where the sick or dead have been.

LEG WEAKNESS.

True leg weakness, the kind that at first affects the legs only, is confined almost wholly to the large breeds and is caused by rapid growth, which increases the weight out of proportion to the strength of the legs. The tendency to this form of the disease may be lessened by feeding bone-forming food. The trouble usually begins when the chicks are between four and five months old. The first symptom of leg weakness is a shaking or trembling of the legs when the chick stands or walks. Generally the appetite is good, even after the patient is unable to walk.

Treatment, to be effectual, should be commenced as soon as the first symptom of weakness is seen, for, after a chick once gets down upon its hocks, it is almost impossible to get it up again. The first thing to be looked after is the food; if the feed has been mostly corn meal, change to shorts and whole wheat, and give a raw egg daily to every two patients. Give milk to drink and give a teaspoonful of bone meal each day to each chick. Keep lime where it is easy of access. For medicine, give half a teaspoonful of Douglass Mixture a day to each chick and twice a day give a half-grain pill of quinine. They should show signs of being better in a week; then give only one pill a day, and, as soon as shaking ceases, leave off the pill and the egg, but continue the bone meal three or four times a week. Let them have the Douglass Mixture for three or four weeks longer. If a week of steady treatment does not produce improvement, it is not advisable to fuss with them any longer. Use the hatchet.

When leg weakness comes on half-grown chicks of the small breeds, give pills and Douglass Mixture as for the large breeds, till they brace up, then feed bone meal, lime and meat right along till carried to marketable age and condition, then market them. Do not keep for breeders.

While fowls are being treated for leg weakness, keep them by themselves, but not in a close coop—they need exercise. To prevent leg weakness, breed from healthy stock, and feed bone meal.

Leg weakness in old fowls is sometimes caused by too high feeding and too little exercise. The fowls get fat and heavy and cannot walk or stand steady. This form of weakness may sometimes be overcome by cutting down the feed and giving the Douglass Mixture, bone meal, and making the fowls scratch. But it really pays better to kill at once and use for the table any fowls that begin to show leg weakness because of overfeeding. Killed then they are perfectly wholesome.

Another form of leg weakness comes from injury to the hock joint, caused by jumping from high roosts. To prevent this, have low roosts. If it occurs before the roosts are made low, put the injured fowl in a well littered coop, without perches, and give victuals and drink, and nature will do the rest, if cared for when first lamed.

Paralysis of the legs is different from other forms of leg weakness, and the best remedy is the hatchet.

Gout may be told from leg weakness, as the legs and feet feel hot, are somewhat swelled and have a more or less inflamed look. It is most common in the Asiatic breeds. Remove the bird to a dry, warm place and give a dose of calomel to open the bowels, after which a half-grain pill of colchicum should be given daily. If the legs and joints are daily well rubbed with sweet oil they will be benefited.

LICE,

though not a disease, are generally classed as such, for a louse-afflicted fowl is as bad off as one that is diseased. It seems hard that fowls should be made to suffer so much inconvenience, and their owners so much loss, because of unthrift through neglect to provide means whereby they may keep themselves free from lice. Cleanliness will do it. When lice have got a foothold in a flock or their quarters, the real work begins, for they must be dislodged or health and profit are gone. The ways to dislodge them are very nearly as various as the parties who find their flocks infested. We give a few of those most commonly used.

The Symptoms of Lice are various. Bowel disease in

summer is one; when chicks are sleepy or drowsy, look out for lice; when fowls refuse to eat, when they look puny or grow slowly; when they die suddenly; when there is a gradual failing or wasting away; when there is constant crying or loss of feathers, look out for lice. Even when houses are kept clean, large body lice may be found on chicks; these come from the adult fowls. A chick will never get these lice unless old fowls are near; that is why brooder chicks grow faster than those under hens. The large lice will kill ducks suddenly. They kill nearly all the young turkeys that die that are not killed by wet and damp. Whenever you notice a sick fowl dusting itself look for lice.

The Dust Bath.—If the house is kept clean, and a dust bath provided, the hens will drive the little mites away, but it is not easy to get rid of the large lice. Finely sifted coal ashes or dry dirt, especially road dust, is excellent. A little sulphur added to the dust bath is a help, as will be a little air-slaked lime. The poultry must have a dust bath, both in summer and winter.

Causes of Lice are not many. Filth is the greatest cause. The mites will breed in the droppings. Rotten nest eggs will cause them. If an egg is broken in a nest and allowed to remain there, there will soon be multitudes of lice. The hen that is sitting breeds them by thousands. They are harbored in cracks and crevices everywhere. They leave their quarters at night and prey upon the fowls, but the large body lice never leave the birds; you have got to look for them and look carefully too. Although these large body lice never leave the fowls of their own accord to find harbor in the house, there should be provisions made for the hens dusting themselves. Also keep the house and surroundings clean and well saturated with kerosene oil. Having provided a dust bath, take each fowl by the legs and dust plenty of Persian insect powder (have it fresh) into the feathers and down. Then grease the heads, throats, legs and vents with a mixture made as follows: Lard, one teacupful; carbolic acid, one-half teaspoonful; crude petroleum, one teaspoonful; oil pennyroyal, one teaspoonful; kerosene, one teaspoonful. Mix well, and use only a few drops on each place.

Never grease the body of a fowl or chick, nor use kerosene undiluted.

A drop of pure lard or oil may be put under the wings; this will kill any lice there.

To get rid of Lice.—Clear everything out of the coop, burn the rubbish, whitewash the sides and roof, pour coal oil into the cracks and crevices. If the floor is of earth or covered with earth, scrape off the top and carry it away out of the reach of the fowls, then sprinkle the floor with air-slaked lime or fresh earth. This done, fumigate the house with burning sulphur. This is done by putting the sulphur in an old kettle, setting fire to it and shutting the house up tight. If the perches and nest boxes are not too numerous or expensive it will be the cleanest thing to burn them and all other material removed from the buildings. If they are not burned, take boiling hot soap-suds and give them a thorough scrubbing. When they are dry and replaced in the building wet them with coal oil. The old nest material should not be used; the new should be sprinkled with snuff, carbolic powder or insect powder. Air the build.ng well before letting the fowls in. Repeat the oiling of perches and nest boxes once a week. Dust insect powder or Scotch snuff into the feathers of the fowls. Give them a good dust bath of sifted coal ashes or road dust to which a liberal supply of sulphur has been added. Keep the house clean and whitewash at least twice a year. This fight persisted in will get the start of the lice.

Sassafras Oil for Lice.—J. W. Crise writes to the Poultry Keeper: "I find sassafras oil used on the roost is more effective than anything I've ever tried. One ounce is sufficient for 100 fowls. This amount put on the roosts will last a long time, and it is said that lice will not stay where it is used, and I know this is true. Sassafras poles for perches would be preferable if convenient."

Tansy as a Remedy.—E. E. Kennicott writes: "There has been considerable talk in the Poultry Keeper in regard to the various ways of exterminating vermin in the poultry house and from the fowls, and I will give my way, which is simple and has always proved successful with me. Gather tansy in the fall, and tack it up around the chicken house, under the roosts, in the bottom of the nests, and on the ground. This will drive away both the body lice and mites. Use lots of it. It is cheap and effectual.''

Crude Petroleum.—H. J. Fisher gives this for a remedy: Procure crude petroleum, and a whitewash brush; go to work and apply the oil plentifully to the inside of the coop, nest and perches. No chicken lice will apply, and those in the vicinity will turn up their toes. Very lousy birds will require a greasing with the oil. Tar paper, sulphur, etc., do not compare with the petroleum. Again, another use of petroleum: For each bird in the flock add from five to ten drops of petroleum to bran mash, feed once or twice daily, and chicken cholera will, as the boys say, skip out. Crude petroleum can be had for from $1 to $1.50 per barrel near oil centers.

Tobacco as a Remedy.—F. H. Putman says: I took two pounds of plug tobacco, soaked it thirty-six hours in three gallons of rain water, occasionally pressing it to obtain all the strength possible from the tobacco, and then turned the water off into a large pail, and with an old whitewash brush I covered the roosts and nest boxes thoroughly with the tobacco juice. I also sprinkled the nests, even under the sitting hens, and in twenty-four hours I could not find a single chicken louse on any old or young chickens that I examined, and am confident it is a sure destroyer of chicken lice. I also had quite a large flock of young Pekin ducks, and was finding one or two dead in the coop each morning. Since using the tobacco I have not lost a single duck, so I think lice was the cause of their dying too.

Bisulphide of Carbon.—A French writer says he drove the lice from his hen-house by tying a few small bottles of bisulphide of carbon to the perches with the stoppers out, leaving the liquid to evaporate. The hens roost over the bottles and the vapor kills the lice. This is what he says: ''The very next day after using it I was agreeably surprised to find that the enemy had left, leaving none but dead and dying behind, and on the following day not a single living insect was to be found, while my birds were sitting quietly on the roosts enjoying an unwontedly peaceful repose. This lasted for twelve days, till the sulphide had evaporated. Twenty-four hours later a fresh invasion of lice put in an appearance under the wings of the birds in the warmest portions of the house where there were no currents of air. I replenished the supply of sulphide, and the next morning only a few of these were remaining. The next morning every trace of vermin had disappeared. Since that time I have personally made a great number of further trials with the sulphide with immediate and absolute success. I should recommend the sulphide of carbon to be put in small medicine vials hung about the pigeon-house or poultry roost. When it has about three parts evaporated the remainder will have acquired a yellowish tinge, and no longer act so completely as before, but if it be shaken up afresh it will suffice to keep the enemy at a distance."

If this is used great care must be taken not to have fire of any kind near the bisulphide, and it seems to us that such fumes strong enough to kill lice on a fowl would be hazardous to the life of the fowl. Furthermore, bisulphide of carbon is heavier than air and sinks instead of rises. We can hardly see how it can be of service as recommended. We give it here with these comments, that our readers may be posted should they see this remedy recommended.

CHOLERA

is by all odds the most contagious and rapidly fatal of all poultry diseases. It attacks turkeys as well as fowls. It is pretty well settled that a microscopic organism or germ, taken into the system through food and drink, causes all the trouble. This germ first affects the blood, then the liver, and throws the whole digestive apparatus out of order. Where the germs generate or are begotten is as yet unknown or at least not revealed to common poultry folk. There is quite a general impression in the air that filth is a good breeding place for these little enemies of poultry life. There are cases where the quarters and runs are clean and cholera takes hold of the flock. Here the germs may have been brought in by neighboring poultry, or the feeding utensils or watering vessels may be filthy. Even where the fowls roost in trees and the droppings are allowed to accumulate, the putrid nature of the mass is revealed when damp or wet comes, and cholera has been known to break out when no other cause could be given but this pile of corruption in the open air.

A. J. Hill, in his "Treatise on Chicken Cholera," says he has sufficient evidence to warrant him in saying that the cause is local. Wherever the disease prevails, right there the cause exists; and there is the place where its cause was generated, unless infectious matter had been introduced by diseased fowls or otherwise.

Symptoms.—The same writer describes the usual symptoms thus: "The fowl has a dejected, sleepy and drooping appearance and does not plume itself; is very thirsty, gapes often and sometimes staggers and falls from weakness; comb and wattles lose their natural color, generally turning pale, but sometimes dark. There is diarrhea, with greenish discharge, or like sulphur and water, afterwards thin and frothy. Prostration follows, the crop fills with mucus and wind, the breathing is heavy and fast, the eyes close and in a few hours the fowls die." There may be slight variations in the symptoms of different fowls, but the peculiar color of the discharge and their frequency are sure indications of the disease. Some fowls will live several days after diarrhea commences, others in apparent good health one day will be dead the next.

Preventives.—Poultry breeders are divided in their opinions as to whether cholera is contagious or not. It is best to be on the safe side and assume that it is contagious, and act accordingly. It is certain that where cleanliness is rigidly maintained and disinfectants are freely used, cholera may be almost, if not entirely, prevented; if introduced into such quarters by strange fowls, or from the premises of neighboring flocks, it can generally be soon eradicated. As an aid in keeping the flocks healthy, we repeat what we have already said in regard to cleanliness and care. Have dry runs, houses free from damp and well ventilated, without drafts. The yards and houses may be kept clean and sweet by using the scraper (hoe or spade) and plenty of fresh soil, fine sand, sifted coal ashes or sawdust.

Whitewash is a great preserver of freshness; it purifies the air, sweetens the premises, and gives a light and cheery aspect to the whole place. Use plenty of whitewash, and use twice or oftener each year.

Do not feed lice, but feed your fowls wholesome food in variety. Use thrifty, vigorous stock for breeding purposes. Have a pen for sick fowls, and when you get new stock hold them in quarantine till they establish the right to be called healthy; then give them the privilege of the place. Keep every fowl busy, and if any are too tired to work or are a little off their feed, reduce their ration, and for awhile give a few drops of Douglass Mixture or tincture of iron, daily, in their drink.

Disinfectants.—When there is any contagious disease among your flock, or in the neighborhood, disinfect your houses, runs and all places frequented by the poultry, at least once a day, until all danger is past. A good. cheap disinfectant is made by adding two ounces of carbolic acid to three quarts of water. Or dissolve three pounds of copperas in five gallons of water; then add half a pint of crude carbolic acid. These may. be applied with a common sprinkling can. Whitewash is a good disinfectant. Lime or ashes will help to sweeten ground that has been fouled by fowls roosting in trees above it.

Remedy.—The surest, quickest way to get rid of cholera is to kill all the sick fowls and burn the remains, or else bury deep, having first covered them with quicklime; then clean up the quarters and burn the matter gathered. When you have got the quarters as clean as possible, close them up tight and put a pound of sulphur in an old kettle, pour on half a pint of alcohol, and have it where you can reach it from the door, set fire to it, close the door, and leave it to burn out. Do not have any fowls in the house during the burning. Wherever the sick fowls have left their droppings, wet the ground thoroughly with the copperas disinfectant, and scatter lime freely. Do this daily as long as the cholera remains upon the premises. Give the flock pulverized charcoal (tablespoonful to pint of food) three or four times a week, and once in every two or three days add five drops of carbolic acid to a quart of water and use it to mix their food with; also, till after the cholera has disappeared, give Douglass Mixture daily in their drink.

A Tested Cure.—A Kentucky subscriber writes me: I want to thank you for your answer to my request as to the trouble with my poultry and for the remedy. After I wrote you and before receiving your reply, I discovered another remedy for the cholera that played such havoc with my fowls. It was such an effective one I want your readers to have the benefit of it. It is as follows: Equal parts of saltpeter, black antimony and sulphur. Mix the powdered sulphur and black antimony thoroughly, say a teaspoonful of each, then mix this dry with the meal or bran whichever is intended for the feed; then dissolve the saltpeter in warm water enough to make the mass the usual consistency for feeding. A teaspoonful of each of the saltpeter, black antimony and sulphur is about the right proportion for a feed for ten hens.

If the hens are too sick to eat put the feed in their mouths with the fingers, and see that it is passed down their throats. When I first tried this remedy there were two of my best hens to all appearances as dead as Hector, they were just breathing and that was all; didn't expect to save them. They were given the remedy in the evening a little after dark; in the morning they were on their feet, and on the third day after were about as well as ever. I feed the mixture occasionally as a preventive. Have not lost a single fowl since commencing to feed it. Have no doubt that it is good for hogs also.

OTHER REMEDIES.

If you do not wish to try the hatchet remedy on the sick fowls separate them from the well fowls. Give the well ones the acid, pulverized charcoal and Douglass Mixture as recommended in "Remedy" above and try some of the following on the sick ones: Calomel and blue mass in two-grain doses may be given twice a day.

Powdered chalk, powdered charcoal, gum camphor,

asafetida and pure carbolic acid, equal parts; mix together and give a teaspoonful in food twice a day to ten fowls.

Half a level teaspoonful of hyphosulphate of soda in as much water as will dissolve it. Give once a day for three days.

Dr. Dickie's Remedy.—Fowls that are too sick to eat should have every four or five hours a pill made as follows: Blue mass sixty grains, pulverized camphor twenty-five grains, cayenne pepper thirty grains, pulverized rhubarb forty-eight grains, laudanum sixty drops. Mix and make into twenty pills. When they have had time to act, give half a teaspoonful of castor oil and ten drops of laudanum to each. Let them drink scalded sour milk, with a gill of Douglass Mixture for every twenty-five head, a day. This treatment ought to change the character of the evacuations and make them darker and more solid. When this happens, and not before, give them alum water or strong white oak bark tea to drink, and no other drink. This will tend to check the discharges.

SOFT-SHELL EGGS

are classed with diseases by most poultrymen. Of course, they are not the disease but the result of a disease, or a lack in the material furnished the hen. It used to be considered that lack of lime was the fault, now the trouble is assigned to overfatness, or a weakness of the egg-producing machinery. Some claim they care not how much lime, oyster shells or egg shells be given a hen, if she is too fat soft-shell eggs double-yolk eggs, infertile eggs—all departures from the normal—will be the result. Those who claim that lime, furnished in the above forms, being insoluble, is of no use, hold that lime should be furnished in oats, wheat, barley, and especially in clover. It is undoubtedly true that a large portion of the good that oyster shells do poultry is the mechanical part they perform in the grinding operations of the gizzard, yet from careful experiments it is shown that there is a certain amount of direct good obtained by their use, and, where they can be obtained at a reasonable cost, it will pay to use them. See "Oyster Shells for Laying Hens."

ANÆMIC POULTRY.

Prof. Woodroffe Hill, of England, says that in the numerous specimens of poultry submitted to him for investigation as to the cause of death, he frequently finds it to be anæmia. The term anæmia signifies poverty or deficiency of blood. In this disease a great diminution in the quantity of red globules or corpuscles takes place; from the normal condition of 130 per 1,000 of blood they are in advanced cases of anæmia reduced as low as fifty per 1,000. The liquor sanguinis, or fluid in which the corpuscles are suspended, is deficient in albumen, and has generally an excess of saline matter. It is important to recognize the gravity and results of anæmia, for I find amongst poultry people it is a condition very frequently passed over, and when otherwise, not very clearly understood; therefore I wish to make this article as plain as possible. It is necessary for the maintenance of health and strength that the food should not only be good and suitable, but properly assimilated after being partaken of—i. e., converted into nutrition—and it will be easily seen that anything affecting the nutritive process must be injurious to the functional activity of the digestive and other organs. This is especially the case with anæmic or poor blood, which fluid under such conditions not only deteriorates the power of the gastric and intestinal glands, but weakens the muscular action of the stomach proper, and its important secondary agent, the gizzard. It will be, therefore, understood that anæmia plays a prominent part in the production of indigestion.

If the reader will pause for a moment to consider the important part in the maintenance of life the red corpuscles of the blood play, remembering they are the agents by which the chemical changes occur in the body, their emission of carbonic acid gas and absorption of oxygen in the lungs, their ceaseless circulatory rounds conveying oxygen to every part of the system, aiding in the removal of effete matter, and constantly building up the body with nutritive elements, he will recognize at once, or should do so, the value of their mission, and the importance of maintaining their standard of strength. In anæmia the center of circulation (the heart) is of necessity weakened, and it is almost needless to say this great force pump requires a full and free supply of healthy blood to enable it to maintain its strength and perform its work properly. The power of contraction and dilation which the heart must continually exercise is strengthened or lessened in accordance with the amount of material the organ is supplied with, and to which it owes its machine-like regularity and muscular energy, and the feeble heart-beat of a poor anæmic little chick very soon stops. Anorexia, or loss of appetite, as associated with anæmia, is the result of the weakened state of the digestive organs, the tone of which being lost, the sense of hunger becomes blunted, and the bird has consequently little or no inclination to feed.

The Causes of anæmia are numerous and not difficult to find. Overcrowding, defective ventilation, stinted light, bad drainage, innutritious and insufficient food, are severally conducive to anæmia, and if the subject be of a weakly constitution they are the more so. Anæmia also follows debilitating disease and hemorrhage. Cellar-kept poultry or those in other dark habitations, soon become anæmic. Note the bleached and colorless shoots of a plant that has sprouted in a dark cellar and compare them with the shoots of a similar plant exposed to heaven's light and breath, or observe the pallid countenance and languid step of an individual who is confined in a crowded, ill-ventilated workshop throughout the day, as contrasted with one whose occupation gives him every chance of imbibing pure, or at any rate fresh air, and you have a true and daily illustration of the effect of these sanitary arrangements, which may be with equal force applied to poultry under similar conditions. Indeed, fresh air and light are as essential to birds of the gallinaceous tribe, for the formation of good blood, as to man. Air must, to maintain health, be renewed not re-used. It is the oxygen which gives color to the blood. Stint the supply of this necessary element and you withdraw the coloring matter and promote the pallid condition characteristic of anæmia. Again, good nutritious food is just as necessary for the production of pure blood and healthy muscle. We may as well try and build a strong substantial house out of bad and weak materials, as expect that blood derived from such a source, and under the circumstances enumerated, will make sound muscle.

Symptoms.—Anæmic poultry generally exhibit considerable muscular prostration, with depression of spirits. The bird has a bloodless look, especially about the eyes. The comb is generally pallid, cold and inclined to lop over. The mouth is white, the tongue particularly so. The limbs are cold, and the thighs sometimes swollen. The skin is unnaturally white and clammy. The bird very often squats or walks languidly about, as though life wasn't worth living. A post-mortem examination reveals general pallor of the muscles and viscera. The tissues are flabby and watery-looking, the liver bleached, and the lungs of a grayish-white color. Anæmic birds are usually emaciated. The eggs (but few) are thin in shell, and pale in yolk. The excretions and secretions are scanty, the plumage lusterless. Indigestion and loss of appetite have already been alluded to. Anæmic poultry is not nutritious food or readily digested, any more than anæmic veal—i. e., where the calf has been frequently bled to produce white meat after slaughter.

Treatment.—To insure a successful issue the causes giving rise to anæmia must be promptly removed, and this should be followed by assisting nature in restoring the deficiency in the color and quality of the blood by those agents which form the necessary constituents of healthy blood. For the former a nutritious diet, with a free allowance of fresh air, sunlight, and ample run should be ordered. Vegetable and mineral tonics, especially the preparations of iron, and, if there be much emaciation, cod-liver oil should be prescribed for the latter. The phosphate of iron is extremely serviceable in anæmia, and when the latter is associated with indigestion I find the greatest benefit from steel and pepsine pilules. In advanced cases the inhalation of oxygen may be had recourse to. Much valuable poultry is lost yearly from anæmia, and yet I know of no condition so easily avoided and remedied, which makes it all the more deplorable that the sacrifice and loss goes on unchecked.

TONIC POWDERS.

No. 1.
Licorice, 2 oz.
Ginger, 2 oz.
Cayenne Pepper, 1 oz.
Anise Seed, ½ oz.
Pimento, 2 oz.
Sulphate of Iron, 1 oz.
Powder and mix.

No. 2.
Cassia Bark, 1½ oz.
Ginger, 5 oz.
Gentian, ½ oz.
Anise Seed, ½ oz.
Carbonate of Iron, 2½ oz.
Powder and mix.

No. 3.
Peruvian Bark, 2 oz.
Citrate of Iron, 1 oz.
Gentian, 1 oz.
Pimento, 2 oz.
Cayenne, 1 oz.
Powder and mix.

No. 4.
Cascarilla Bark, 2 oz.
Anise Seed, ½ oz.
Pimento 1 oz.
Moth Dust, 2 oz.
Carbonate of Iron, 1 oz.
Powder and mix.

No. 1 is especially recommended in case of sudden colds. No. 2 is an excellent tonic in wet or cold weather, or for young turkeys. No. 3 is helpful in overcoming the ill effects of the show pen No. 4 may be used when a continuous tonic is required, as when fitting birds for exhibition. This may be mixed with sugar, one part of the powder to three parts of refined sugar. This is relished by the birds. When using either of the tonics otherwise, enough should be added to soft food to give it a slight flavor of the tonic and no more, except in special cases, then give each bird what will lie on a dime.

ROUP

is the bane of the poultry yard, and if we except cholera there is no disease so troublesome or offensive. After it has run to a certain stage it is contagious. If neglected it is fatal. When roup gets into a flock their quarters should be thoroughly cleaned and disinfected. Use a tablespoonful of carbolic acid to a quart of water to disinfect. Sprinkle freely on floor and sides of building. Wash the perches, feed boxes and drinking vessels.

Cause.—The chief cause of roup is a neglected cold. The cold may be from exposure to drafts, wet, damp roosting places, or to too sudden a change from overheated houses to cooler quarters. Fowls that have been stimulated by pepper, or egg foods that are composed largely of pepper, and then exposed to wet and cold, will often catch a cold. A little pepper may be a good thing, but too much of even a good thing is bad.

Prevention is better than cure, so the best plan is to guard against all the causes, then if the disease gets among the fowls in your neighborhood use carbolic acid, charcoal and Douglass Mixture as recommended to prevent cholera. If, after all efforts, the disease gets a foothold in your flock it will be of a milder type and more easily handled than if the quarters are damp and filthy, for notwithstanding that fowls sometimes get along amid filthy surroundings without roup, it is a fact that roup thrives and develops the most malignant form in damp and dirty quarters. Like diphtheria in the human family, till the disease appears in a locality the filthy places and the clean are alike exempt, but after its appearance the most filthy surroundings give the best aid to its development.

Symptoms.—The first symptoms of roup—hoarseness, sneezing, and a slight running at the nostrils—are the same as those of a common cold. In the second stage of the disease the discharge from the nostrils thickens and becomes very offensive, and the eyes and head are affected more or less. In the third and last stage the head swells, ulcers form in the mouth and throat, and sometimes around the eyes, the appetite fails, the comb turns black, and the fowl dies. When the roup first makes its appearance in a flock, while it is still in the first stage, is the time to handle it easily and surely. In the beginning the symptoms are identical with those of catarrh, but the discharge soon commences to thicken and fill up the nostrils, the eyelids and face become swollen from the accumulation of mucus, which now gives out an offensive odor, air bubbles appear in the corners of the eyes and in the throat, and in a few days the bird, unless relieved, dies from suffocation. When the disease assumes this aggravated form, it becomes highly contagious; therefore, no time should be lost, but the affected bird should be removed from its companions, and thus prevent the communication of the disease.

Names.—The term roup covers a multitude of ills. It is known by different names, such as sore head, sore throat, inflamed eyes, swelled head, cancer, catarrh, pustulated nostrils, but in each and every case roup would cover the symptoms, and the remedies employed for it would alleviate or cure them.

Remedy.—Disinfect the quarters by cleaning as thoroughly as possible, then shut up tight, put a pound of sulphur in an old iron kettle, pour on a half a pint of alcohol, set it in the house where you can reach it from the door, hold your nose with one hand, set fire to the alcohol, shut the door and leave. Do this after the sick ones have been separated from the well ones. Give the sick ones a dessert-spoonful of castor oil at night, and for a week feed chiefly on cooked food, with daily doses of charcoal, Douglass Mixture and acid. The well ones may be given the charcoal and Douglass Mixture for a week also. If any of the sick ones have ulcers in the throat dust them twice a day with pulverized chlorate of potash. If the fowls are so bad that the nostrils are clogged with matter, the head swells, eyes are closed and ulcers are in the throat, use the hatchet at once and burn or bury deep the "remains."

Another Remedy.—Although many claim a sure cure for this disease, treatment of the sick ones is very unsatisfactory. Remove the sick ones and thoroughly clean and disinfect the chicken premises. Scatter lime and carbolic acid freely about the walls and floor of the house. Give a warm, dry place to stay. Rub the throat of sick ones with coal oil and camphor. Swab the throat with kerosene.

The Hatchet Remedy will give the best satisfaction when the disease has reached the third stage.

P. H. Jacobs gives the following treatment: As soon as hoarse breathing is noticed, and especially when the bird is suffering from severe hoarseness and seems to be in danger of choking, put it in a large box and set fire to a mixture composed of a tablespoonful each of pine tar and turpentine, with a pinch of sulphur and a few drops of carbolic acid.

Keep the bird in the box until nearly suffocated, when the breathing will at once become easier and the disease more readily submit to treatment. Burn some of the same ingredients every night, in the poultry house, after the birds are on the roost. If there are many sick fowls place several of them in a large box, barrel or hogshead, and submit them to the fumes of the mixture together, or the whole flock may be so treated every evening when shut up in the poultry house. It detaches the phlegm and membrane and causes the matter to be thrown off. Having done this for the croup form of roup, inject two drops, twice a day, in each nostril, of the following: Bromo-chloralum and water, equal parts.

Injections for the Nostrils. — Should the bromo-chloralum fail to give relief, mix together a tablespoonful of kerosene oil, the same of warm lard, and add ten drops of carbolic acid. Inject two drops in each nostril once a day, using a small glass eye-syringe, or a sewing-machine oil-can. Another excellent injection is Labarraque's solution of chlorinated soda, mixed with twice its quantity of water, using two or three drops, twice a day, in each nostril.

Swelled Head and Sore Eyes. — When the eyes are sore, and closed, and the head swelled, bathe the eyes with a warm solution, once daily, made by dissolving a teaspoonful of boracic acid in a gill of water, using a soft sponge for that purpose. Once a day, also, anoint the head and eyes (closed) with ten drops carbolic acid in a tablespoonful of glycerine. The following is also recommended by some as a wash: It is to use eight grains sulphate of copper (blue vitriol) and six drops solution of carbolic acid for each fluid ounce of water. Apply this wash two or three times a day, by means of a camel's-hair pencil, to the face, taking care not to injure the sight by allowing it in the eyes; brush the inside of the mouth and throat, and inject it by means of a small syringe into the nostrils. As the disease abates reduce the frequency of the application and the strength of solution. A wash which may be used in place of the above, and in the same manner, and without fear of injury to the eyes, is the solution of chlorinated soda (Labarraque's solution) diluted with four times its bulk of water. In this and all other diseases, much is gained by taking the case in hand at the earliest stage.

How to Treat.—After the disease becomes contagious, first thoroughly disinfect the entire premises, and use bromo-chloralum and dilute it one-half. Inject it up the nostrils once a day with a small syringe or a sewing-machine oil-can. Add sixty grains of bromide of potassium to each quart of drinking water. Burn a mixture of wood tar, turpentine, sulphur and carbolic acid in the poultry house at night, after the fowls have gone on the roost, until they are nearly suffocated, and repeat every evening. Bathe the heads with warm water, adding ten drops of carbolic acid to each gill of water. Above all things, avoid cracks, crevices or drafts, especially from ventilators at the top. The head and throat may be greased once slightly (no more) with an ointment composed of lard, kerosene and turpentine, equal parts. For rattles and canker throat and mouth, use one ounce chlorate of potash in a pint bottle, sixty drops tincture iron, twenty drops carbolic acid and twenty grains bromide potash. Fill with water and give one-half teaspoonful night and morning.

Tonics.—Give these in the soft feed, morning and night. Take one dram Peruvian bark, one dram gentian, twenty grains bromide potassium, ten grains pulverized copperas, one dram salt and ten grains red pepper. Give a teaspoonful in the soft food for five fowls.

An Alum Remedy.—Mrs. Johanna Hunter, Kansas, says: "My chickens were dying of roup. I tried many things but all failed. I then put alum in the water that I gave them to drink, made it very strong for two weeks, and then have given it once a week. I have 200 hens; have got over eighty dozen eggs in the month of January; have lost but very few chickens since giving the alum."

The Canker Form is thus described by a subscriber to Poultry Keeper: "The first thing I noticed was a sore on the outside of the face, a little back of the opening of the bill. In opening the mouth, I found one side covered with thick canker, and the whole side of the head is now sore, and blotches down the throat. Is this roup?" Yes, it is roup, and when lumps and sores appear it is verging on the scrofulous. It is useless to attempt to save such a bird, as the labor required would be too great, and the disease may spread. As a remedy a tablespoonful of chlorate of potash in each quart of drinking water, with the anointing of the face with a few drops of a mixture of one part spirits turpentine, and three of sweet oil, would be excellent.

Eye Ointment.—When the eyes are covered with matter they may be anointed with a mixture of one part spirits turpentine and three parts sweet oil.

Soap Remedy.—Here is a receipt for the roup which has never failed me for the last fifteen years. It may be of use to some of your readers. Take common soap, scrape off with a knife from the bar as much as you require, and work into the same as much red pepper as it will take. Give two pills the size of a hazelnut. If one dose don't fix them, a second dose the next day will.—F. G. Lee.

EGG-BOUND

is caused by the hen being too fat, by the attentions of a heavy cock, by jumping from a high roost, or by injury of some kind, but overfeeding is the main cause. It may be known by the appearance of the hen from the rear. If the egg gets broken it will usually prove fatal with the hen, and for that reason great care should be exercised in treating. The first step is to oil the vent with pure olive oil; also inject a little into the egg passage. If that does not give relief within an hour repeat, and in addition bathe the parts with something warm and moist. The food should be soft, and but a small quantity given until the egg passes. If an ordinary fowl we advise killing for the table before fever sets in. If the hen is valuable it may pay to give her careful attention until relieved. The following has been recommended: One grain calomel, one twelfth of a grain of tartar emetic and a quarter of a grain of opium, made into a pill, and administered every four hours. In the first pill the quantity of calomel and opium may be doubled. The chances are small that a hen which has become egg-bound at any time will be of any value afterwards as a layer.

CROP-BOUND.

When a fowl's crop is hard and about twice as large as it

ought to be there is something that prevents the contents from passing into the gizzard. This trouble is called crop-bound, and to unbind it pour some warm water down the throat and then carefully knead the crop until the contents are somewhat softened, then hold the fowl's head down, open the beak and work at the crop a little longer. After this, give a tablespoonful of castor oil and shut the fowl up for ten or twelve hours without food. At the end of that time if the crop is not empty, or partly so, cut it open and remove the contents. Make an inch and a half opening in the upper part of the crop. Have a small sharp knife, and be careful not to cut any of the large blood vessels. After the removal, oil your finger (use sweet oil) and pass it carefully as far as possible down the passage toward the gizzard so as to be sure that there is a clear track for future meals. Take two or three stitches in the crop, also in the outer skin, using care that you do not sew the one to the other; use silk thread. Shut the fowl by itself and feed lightly on soft cooked food for ten or twelve days, giving no drink for two days after the operation. Some prefer warm lard to water. Sweet oil may be used.

SCABBY LEGS

are due to minute parasites too small to be seen, but which rapidly multiply. Scabby legs make the bird a filthy, disagreeable object. To cure it is easy if taken in hand when it begins to make its appearance, as an ointment composed of one part coal oil and two parts lard will clear it entirely off, but when the legs become thickly covered with heavy scales or shales, some work must be done. First scrape away as much of the scale as possible, and grease the legs, from the thighs to the toes, with the following ointment: Gas tar, one gill; lard, one gill; carbolic acid, one teaspoonful; coal oil, one tablespoonful. Wash and dry the legs after scraping them, and rub the mixture on well. Do this every week, repeating the rubbing and scraping, and as the parasite which causes the disease will soon succumb to the ointment, the scale will gradually come off. The best plan is to use the ointment early and often, as the disease is contagious.

Another Remedy.—The above remedy is the one prescribed by P. H. Jacobs, editor of the Poultry Keeper. The following is one given by Fannie Field, the noted poultry keeper of Massachusetts. Dip the fowl's legs into coal oil and hold them there until the oil has had time to penetrate beneath the scales and kill the parasites that do the mischief. These applications, with intervals of a day or two between, will generally effect a cure. The scales will loosen and fall off. Do not try to hasten their departure by rubbing or scraping them off. Rub the legs carefully every day with melted lard or sweet oil until they are smooth and well. Scaly legs or scabby legs are contagious. That is, the parasites go from one fowl to another till the whole flock is afflicted unless their "run" is stopped; hence the first affected fowl noticed should be at once dealt with.

To us the Fannie Field remedy seems the simplest.

FEATHER-EATING

is a vice rather than a disease. Like all bad things, it contaminates all within its reach, so when a fowl is noticed doing the unclean thing the surest way to stop the ill effects of this bad example is to use the hatchet and put the offender in the pot. If a valuable bird it may be broken of the habit by the use of a bridle which can be bought for a few cents.

A Remedy recommended by some is this: Make an ointment of sulphur, kerosene, lard and carbolic acid. Anoint that part of the plumage that is being pulled out, and the offender, not relishing this "sauce," may soon stop its offense. One poultryman fed his flock feathers and they soon got disgusted with such fare and behaved themselves. It is very seldom a busy flock has any feather-eaters among them. It is idleness that begets the evil. Keep the flock scratching, hustling.

INDIGESTION.

When a bird walks lazily about, with little appetite, hardly touching ordinary food, while droppings are scanty and unhealthy in character, it is pretty certain something is wrong with its digestion. The causes vary, but the treatment should be the same. Give daily five grains of rhubarb, changed every fourth day to one grain of calomel. Restrict the diet to a small allowance twice daily of soft, well cooked food. Let the fowl drink after each meal, then remove the water vessel. A little finely cut green grass may be given several times a day.

ENLARGEMENT OF THE LIVER.

Very often simple indigestion neglected may terminate in serious enlargements or other disorders of the liver. Overfeeding—especially with highly seasoned foods—or other errors in diet, and lack of exercise may be the cause. The symptoms are laziness and lack of appetite, with a sickly, yellowish look about the head and comb. If there is much enlargement of the liver, treatment is of little avail and the hatchet would be the best thing to use; but as there is no way of telling the size of the liver until after death, there is a desire to put off death as long as possible. The rations must be cut down. Give a grain of calomel every other day for a week, give as much range as possible, and feed green onions.

LIVER COMPLAINT.

In the Iowa Homestead a reader says: One of my chickens died. It stood around, with its head drawn back to the wings. The head looked pale, and of a yellowish tint. The bird grew light to about half its natural weight. The crop was entirely empty, but the stomach was as hard as a rubber ball, and on opening it I found it filled with gravel and small straws. It had an inflamed look. The inside skin was loose from the outer part, about two-thirds around. The droppings were yellow and white, and thin, as in dysentery. Is the disease contagious, and can you suggest a remedy?

Part of the symptoms point to the common trouble of "growing light," a liver complaint. Probably you noticed when you dissected the bird, that the liver had an unnatural color, and a rotten or cheesy look. When alive, a fowl suffering from this complaint has a regular jaundice and bilious look, with diarrhea at one time and costiveness at another time. There is no positive remedy after the trouble once seats itself. In the early stages, when the bird seems mopish, and the blood seems to leave the comb and wattles, the disease can be checked by giving a family liver pill. Repeat the dose in a day or two. Remove the bird to separate quarters, and feed a warm mash of bran and a little corn meal or middlings. Give a little whole wheat at night. A little condition powder added to the morning mash will greatly aid a cure. Give plenty of green food. Cabbage hung up in the hen-house furnishes excellent greens for fowls in winter. The disease "growing light," is not contagious, but at the same time, the sick birds should be removed to warm and dry quarters.

BUMBLEFOOT REMEDIES.

A Poultry Keeper reader says to the man that complains of bumblefoot in heavy chicks: "Don't let them roost at all, but throw some straw on the floor of the house and let them sit on it, turn it over once in a while and mix in the droppings; when it becomes foul mix with stable manure and save it for the garden. For bumblefoot use an ointment: Tincture iodine, one ounce; oil origanum, one ounce; lard, one half pound. Mix with lard, warm first; don't leave near a flame as it is very volatile."—J. McKenzie, Pennsylvania.

Take the bird to the cellar or some warm house, if in winter, then prepare a poultice made of bread and milk. Take a dish of warm water and place the bird's foot in it for a few minutes. After soaking it wash the foot clean and wipe dry, then place the bird between your knees, take the foot in your left hand, and with a sharp knife cut across the bumble down to the bone. Then make another cut across the swelling, making the letter X, then press the matter out. It will be a thick, cheesy substance. Now apply the poultice, first sprinkling on a little pepper. Leave the poultice on for twelve hours, then change by putting on a new one as before. This may remain on for twenty-four hours, then take it off and wash the foot with warm milk and water, and apply a cloth smeared with pine tar. Leave that on for one week, and when it comes off the foot will be well.—"H. B." Iowa.

DEBILITY

in chicks is often caused by rapid growth of feathers. Feed plenty of meat, give other food in variety four or five times a day. Give of Condition Powder No. 1 (page 134), about half a teaspoonful to each ten quarts of soft food. Also, give a teaspoonful of the tincture of iron to each gallon of water.

WORMS.

Sometimes worms are at the bottom of the "out of order" condition of the poultry. If worms are noticed in the evacuations, there need be no doubt of the trouble, and the fowls should each be given a piece of camphor gum the size of a marrowfat pea. Twelve hours after give another dose of castor oil (teaspoonful) and put a teaspoonful of sulphur and a tablespoonful of powdered charcoal in each pint of the soft food two or three times a week. If there are deaths in the flock from unknown causes, it often pays to examine the dead fowls. If worms are found treat for them. Occasionally "crop-bound" may be caused by worms, or worms may be present in connection with that disorder, if they do not cause it. Not long ago a person wrote to me: "My old turkeys dump around for two or three days with some matter running from their beaks, while they try to swallow. One of them died. I opened it, and upon examination found its craw full, and, oh! so sour, and just lined with little white worms." In such cases we would advise treatment as for crop-bound, followed by that recommended above for worms.

SWELLED EYES

are due to drafts or sudden changes in weather. Wash the heads with warm water, and touch a drop of glycerine to the eyelids.

CHILLED CHICKS.

When young chicks are caught in a shower, fall into the swill barrel or wander through dewy grass and get "chilled to death" there is sometimes life left but it needs to be warmed up or else it will soon go out. If the chick is still able to stand up, drying it off well with warm flannel and then placing in a warm place—the oven of the kitchen stove is the most common place on the farm where brooders are not had—will usually bring the little fellow back to life and actively. When the patient is stiff and cold more heroic measures are needed. Take the chick by the beak and both legs and plunge it into water at 120° at least. Keep the nostrils and eyes out but let all the rest go under. As the cold body cools the water off, add more hot water to keep up the temperature. If he begins to kick and struggle do not treat him harshly, but soon remove him and dry off as mentioned above and give him a warm place, well wrapped in warm flannel. This treatment will not bring a dead chick to life, but it will cause many a chick to live that would otherwise stay "dead," when cold and stiff from being chilled.

CONDITION POWDER NO. 1.

Carbonate of iron, one ounce; anise seed, two ounces; powdered ginger, six ounces; mustard, one ounce; fine salt, two ounces; sulphur, two ounces; licorice, four ounces; powdered charcoal, fourteen ounces. Powder and mix thoroughly. Keep in a tight vessel, fruit-can for instance. A teaspoonful of this to ten quarts of soft food daily will often be of service in keeping the flock in prime condition.

CONDITION POWDER NO. 2.

This powder is more in the way of an appetizer and general invigorator than a medicine, and may be given daily with good results. The amount to be fed is a tablespoonful to five hens. Take two pounds linseed meal, four ounces phosphate of soda, two ounces chalk, four ounces gentian, one ounce ginger, four ounces charcoal, one ounce salt. Have all nicely powdered and mix thoroughly.

DOUGLASS MIXTURE,

the tonic most often recommended for poultry, is made as follows: Dissolve a pound of copperas in two gallons of water; then add two ounces of sulphuric acid. Put in stone jug and keep well corked. The dose is a tablespoonful to each quart of drinking water. When handling the sulphuric acid be careful, as it is poisonous.

CHAPTER V.

TURKEYS, DUCKS AND GEESE.

RAISING TURKEYS.

If one intends to raise turkeys and has not a large range for them, where there is no danger of their continually bothering the neighbors, he had better turn his attention to something else, as turkeys are naturally of a very wandering disposition and can not be successfully raised in confinement. If you can keep turkeys without trespassing on the rights of others, you will find them a profitable adjunct to general farming. Many farmers, farmers' wives and daughters would find it more profitable and really no more work than raising chickens. But there is need of patience and perseverance; do not be disappointed if over half the young turks die the first year—probably more will live the second.

WOMEN AS TURKEY-RAISERS

are usually much more successful than men, and especially those who have brought up families, since, for the first few days of their lives, the poults need nearly as close attention and care as babies.

When starting in do not begin on too large a scale, then the failure will not be so great financially. A few sittings of eggs may be bought of a reliable dealer.

A better way may be to get a gobbler and two or three hens of pure breed. If there are good common hens in the neighborhood and the capital is very limited, a gobbler alone might be purchased; but many consider it cheaper in the long run to have all thoroughbreds from the start. Do not buy too heavy a one if to be mated with common hens, as he may injure them. It is held by some that the birds should be two years old, as "yearlings" are not fully matured and their offspring would be weak. Others claim that a yearling gobbler and hens two years old give as good poults as any matings.

The idea that turkey eggs will not hatch well after transportation or under hens is a mistaken one. Turkey eggs fresh and fertile, from healthy breeding stock, and properly packed, will bear shipping just as well as hen's eggs and will hatch just as well under a hen as they would under a turkey. In fact, some people find that during the first few weeks after hatching, the poults are more easily cared for with a hen than with a turkey.

Before getting your turkey eggs to set make sure of your sitter; have a steady, reliable hen, one which can be relied upon to stick to her job until it is finished, and have her sit where the other hens cannot interfere with her.

Some breeders give seven turkey eggs for a sitting, others nine, others thirteen, but seven eggs are enough for one hen. Whenever it is possible to do so, set turkey eggs outdoors on the ground. By making the nest in a bottomless coop and placing a lath or wire pen in front of the coop where you can feed and water the sitter, the hen is secured from all annoyance by other fowls and all danger from foxes and skunks by night.

Dust the hen well with insect powder to begin with, and again in ten days, and yet again three days before hatching time. Beyond this and the regular daily feeding and watering, let the hen run the business herself when she is sitting on the ground. But if for any reason she sits elsewhere it will be well to sprinkle the eggs with tepid water daily during the last ten days of incubation. Do this just at dusk, carefully lifting the hen from the nest for that purpose.

The gobblers may be kept until four years old and the hens until five. If many hens are kept some of the best should be selected each year to keep up the stock; when the gobbler grows old get a new one not related to him, as turkeys very quickly show the effects of inbreeding.

One gobbler is sufficient for a dozen hens; more have sometimes been kept with one, but it is not advisable. If he is very large and tears the hen's back in trying to keep his hold, cut off the inside toe nails with a sharp knife.

STARTING A FLOCK.

If turkey fowls are purchased, when laying time arrives, nesting places should be prepared for them—old broken barrels "accidentally" cast in an out-of-the-way place, a heap of brush or a few loose boards "dumped" into a fence corner will just suit the hens. Turkeys want to hunt up their own nests and may wander off where the eggs will be hard to find, but if they are kept tame by kind treatment and convenient places fixed for them, there will not likely be much trouble in getting the eggs. While the weather is cold or damp they should be gathered every day and kept in a cool, dry place—a few porcelain or other nest-eggs being left in their stead.

The first eggs may be set under hens, giving seven to nine apiece; if the hen cannot be set on the ground a sod should be put in bottom of nest. After being set a week, if two or three hen's eggs are added, the young chicks will quietly aid keeping the turks from wandering too far, and thus they will be more likely to get home at night. After the weather becomes warm the eggs may be left in the turkey's nest and she will sit when she gets ready; if there are not about twenty eggs then, more may be given her. If hens are set at same time, the poults may all be given to the turkey on hatching.

Twenty-eight days is the time required for turkey eggs to hatch, and the turkey should always sit upon the ground.

If two or more should want to sit at the same time and one can cover all the eggs, give them to her and break the other up. She may begin laying a little later, and with care the late hatched turks will prove profitable. At any rate the gobbler will be more contented; if all the hens are sitting he should be shut up, as he may disturb them and the eggs be lost, or perhaps he will pay his respects to the chickens, and disaster will be sure to follow.

When sitting the hens should not be taken off to feed, but food and water should always be in reach when they come off, which may not be more than once in three days. Sprinkling the eggs with tepid water will be beneficial. Before the day of hatching arrives the hen and nest should be carefully treated to insect powder, as lice are almost certair death to newly hatched turkeys. When hatching, the hen should be left entirely undisturbed, but a coop should be close at hand to receive her and her brood, and she should be carefully watched, as she will seize the first opportunity to lead them off into damp grass where many will perish. Better than to watch her, will be to place a lath fence or wire netting about the nest.

Do not feed the poults any the first day; on the second give them, five times a day, old wheat bread moistened with milk, eggs boiled until they crumble easily, and "Dutch cheese." Give fresh milk for drink. Continue this food a couple of weeks, then make a "cake" of two parts corn meal and one of shorts, season with salt and pepper, and use baking powder to raise it; finely chopped meat will serve well for "raisins." Bake this thoroughly and feed dry, slightly moistening the crust. Give sour milk to drink if easily ob. tainable. Change gradually from one class of food to the other.

If the eggs and milk for the cheese are not at hand, give bread crumbs for a few days, and then the "cake" described above. If they have plenty of sour milk they will not need meat as much as though they had no milk. Feed everything as dry as possible, and don't give more than they will eat at a time.

After they get big enough to catch insects and pick up food for themselves they do not need to be fed so often; night and morning is sufficient if foraging is good. Always give them a little grain at night to keep them in the habit of coming home to roost, and if their crops are not full, fill them.

CARE AFTER HATCHING.

For the first few days after hatching, the hens should be confined in large roomy coops with dry floors; have little yards attached to keep the poults from running away. Then for another week or so keep the mother confined, but on pleasant days after the dew is off let the poults go in and out of the pen at will. They will not wander far enough to tire them out, as they might if the hen were given full liberty with them. After that on pleasant days when the dew is off the grass the hens and broods may be given their liberty:

they should be kept shut up on all damp or rainy days. If the damp weather continues the coops should be moved into the barn or some other place where it is dry and the young "turks" can run about. Have dry earth or sand on the floor. If the business is carried on to any great extent it would pay to have a turkey shed; this is best made with a shingled or tar-papered roof and three closed sides, the front facing south and consisting of large folding doors with windows in them, so that when the sun shines the whole may be open, but yet there may be plenty of light when it rains; the floor should be of sand. Here the coops can be placed and the poults have their liberty.

If turkeys are not kept out of the damp and still be able to run about, the majority will be most certain to die, for turkeys must be kept dry and must have exercise. This is why so many fail to raise turkeys—they do not keep them dry during the early weeks of their lives. Of course where one has a dozen or more broods to look after it would be a big job to get them all to shelter if a sudden shower should come up while the poults are out in the fields; but one must, as they can stand but little wetting, either of dew or rain, until pretty well feathered, and some of them would be drowned, or chilled beyond hope. So where many are raised it is advisable to keep the mothers confined to the coop and run until the little ones are about six weeks old, moving the coops and pens often enough to keep everything clean. This shed can be used as a roosting place in winter and for confining the birds at any other time.

If the turkeys do not return from their foraging expeditions at evening, they should be looked up immediately, as the dew may kill them or something carry them off during the night.

The poults should be watched at all times for lice, and as a preventive is better than a cure, dust them frequently with insect powder, or brush them with a feather dipped in tincture of iodine; don't use kerosene or sulphur.

If at any time they are caught in a rain or escape into the damp grass and get chilled, wrap them up and put by the kitchen fire immediately. If very far gone hold in a pail of warm water until they are fully "waked up," then wipe off and wrap up.

When they become fully feathered and the red about their heads begins to show, turkeys need very little attention except to see that they get to "bed" properly and have had plenty to eat. To make large turkeys for Thanksgiving they must not go hungry for a single day.

Some two months before you wish to market the first of your turkeys, begin to feed them a mash of cooked vegetables stirred stiff with bran and corn meal in the morning and give buckwheat and whole corn at night. Give them plenty of good water; also pulverized charcoal mixed in morning food twice a week. If insects are scarce give meat.

About two weeks before marketing shut up those that you want to keep over or until later in the season, and let those to be fattened have free range (turkeys often grow poor in confinement with heaps of food in front of them) and feed them three or four times a day with potatoes cooked with corn meal, or corn meal mush made with milk, and at night, give whole corn; some other grain might do for a change once or twice. Do not feed more than they will eat up clean, and let the meals be at as long intervals as daylight will allow. If the food is dropped by the handful so that they scramble for it, they will eat much more than if it is dumped on the ground before they are called up.

Turkeys reared and fattened in this way will be a profitable investment, though it may appear before attempting, or on first trial, to be otherwise.

Turkeys are subject to the common diseases that afflict fowls if they are surrounded by the same conditions. The same general treatment will apply to them as to other poultry. The hatchet is the best and quickest remedy; if other treatment is desired consult the chapter on "Diseases of Poultry." The chapter on "Dressing and Shipping" tells how to dress and ship.

RAISING DUCKS.

In many respects ducks are more easily raised than any other kind of poultry, and it is surprising that so many poultry raisers are without these profitable birds as an addition to their business.

It is not necessary to have a large body of water, or even a small creek, to successfully raise ducks. Mr. James Rankin, who raises thousands of ducklings yearly, has found, after careful experiments, that he can rear them without water, except for drinking purposes, more profitably than with it; as when frequenting bodies of water, a greater or less number are always caught by skunks, minks, turtles and other animals, and many get lost in mire and mud. He also keeps his ducks in limited runs.

It is a popular belief that water for bathing is necessary to secure a good proportion of fertile eggs, but this is disapproved by some who have thoroughly tried it.

Ducks can stand colder weather than chickens, but in the northern portions of our country should have a good shelter —a low shed with tight roof, open to the south, will do, but it would be much better to have it boarded up and plenty of windows inserted; gravel is all the floor needed.

One drake to five or six ducks is sufficient. One drake has been kept with seven or eight ducks with good results.

SPRING DUCKS.

If you are situated convenient to a good market, the most profitable way is to hatch the ducks early and turn off when about ten weeks old, for they always command a high price in May and June. The old ducks should be kept in good condition, but not too fat, throughout the fall and winter; one of the best feeds for this purpose is equal parts of corn meal and bran boiled with turnips or potatoes with beef scraps added. Give them two meals of this a day, and for the third, wheat, oats and corn. The corn should be cracked, as ducks do not digest hard food as well as chickens. They should begin to lay about January 1, and then the proportion of corn meal and beef should be increased.

As a duck often lays fifty eggs or more, the first ones should be set under hens and, for that matter, all of them, since ducks as a rule are not very good incubators. The best nest is made by using a sod at the bottom and covering with an inch of chopped straw. During the last two weeks of incubation the eggs should be sprinkled with tepid water, when the hens are off, as otherwise the ducklings may not succeed in getting out, as the shells are quite tough. Many recommend removing the ducks as soon as hatched and placing them in a warm place to dry off, for if left in the nest until all are hatched the older ones are apt to knock the empty shells down over the other eggs and effectually imprison the occupants, or smother the newly hatched duckling by crawling on top of it.

There is this difference between young chicks and ducklings when hatching time comes. The chick a few hours after pipping the shell will be out taking in its surroundiugs. Whereas the duckling, which will be ready about the twenty-fourth day to break the shell, will lie forty-eight hours before it is ready to come out. At this time it is necessary to keep some watch on the hatching, as the duckling is apt to get smothered if the pipped side of the egg gets turned down. Also the shells should be removed from the nest or incubator, to prevent them getting crowded over the pipped eggs, and thereby smothering the little fellows that are seeking air. If the shell of an egg is pipped, but the inside skin or membrane is not broken, it will be well to make a small opening through it with a blunt needle. Be careful not to prick the duckling. Very often ducklings have to be helped from the shells, but do not do this till they are kicking around lively; if you do they are apt to bleed to death.

When taken from the nest, place the hen in a coop with a little pen around it to keep the ducklings "at home." This pen is easily made by setting up six-inch boards and driving stakes to hold them in place. Do not give the ducklings anything to eat during the first twenty-four hours of their lives, then feed every two hours, hard boiled eggs with bread crumbs worked in. Do not boil the eggs too hard; just so the yolk is sticky is considered best by most successful breeders. If you use a tester and remove the infertile eggs from the nests, keep them, for they will be just the thing for early feeding. Be sure to have plenty of water in shallow tins so that the ducklings can not get themselves wet; but they eat

so rapidly that they will choke themselves if they do not
have an abundance to drink.

FEEDING DUCKLINGS.

If you have plenty of eggs continue feeding them, mixed
with bread for a week or more, then gradually change to
corn meal and bran or middlings half and half, cooked in
sour milk if easily obtained; if not use water and baking
powder. Always put some meat in the mixture. Feed this
quite dry four times a day and give chopped grass or other
"greens;" if these can not be secured, cut up clover hay and
steep in hot water. Ducks to do well must have plenty of
green food. Do not give them milk to drink as they will be
sure to daub themselves with it and get their down pulled
out. A little sand is often added to the food to help digestion. The ducklings should be kept from the water until
they are feathered, as their down is no protection and they
are easily chilled, with fatal results. They should be fed all
they will eat up clean, and should be kept growing rapidly.

When six weeks old, three meals a day are sufficient; the
proportion of meat and corn meal should continue to be increased until at eight weeks their feed will be about three-
quarters meal, the remainder being bran or middlings (or
better both), meat and green stuff.

With this treatment the ducklings at ten weeks of age
should weigh about five pounds each, if of the larger breeds
—the Pekin is considered the best, as it grows very rapidly,
and the pin-feathers do not show as much as those of the
dark-colored varieties.

Always save your breeding stock from the best of your
earlier hatchings, if you wish to get strong, early ducks
another year. Do not crowd these with fattening foods.

LATE DUCKLINGS.

If there is no convenient market for "green" ducks, as
they are called when fattened young, or for any other reason
it seems best to keep them until fall, it is not necessary to
hatch so early or crowd them so, and this lessens the expense of raising, but it still more lessens the profit.

If ducks are raised in large numbers it will pay to have

an incubator and brooder, and thus spare the worry and loss occasioned by fickle hens not wishing to sit at the proper time. Chicks may be hatched in December and January, and in this way the incubators and brooders can be pressed into doing double duty. The chickens grow more slowly than the ducks, and will be ready to turn off about the same time.

PICKING DUCKS.

Duck feathers always bring a fair price, especially white ones, and should be saved when dressing the ducks, if they are sold dressed; if not sold dressed do not pick just before selling. The amount received for the feathers ought to pay for the dressing.

The breeding ducks may be picked several times a year, generally four to six. Do not pick until the feathers are "ripe," which can be told by pulling a few from different parts of the bodies of several birds. If they come out easily, without any bloody fluid in the quill, they are all right and should be "picked" or many will be lost. In picking pull only a few feathers at a time by taking between the thumb and forefinger and giving a quick, downward jerk. Do not pull the bunch of long, coarse feathers under each wing.

Before you begin picking, tie the duck's legs together with a bit of listing or other soft cloth and if the duck is inclined to object to the picking by thrusts with the bill, slip an old stocking or something of the sort over its head. Use no unnecessary harshness with any of the birds and be especially careful with laying ducks. Sitting ducks and those that are soon to be set should not be picked. In hot weather much of the down may be taken from the drakes. Do not take any in cold weather.

In handling ducks do not lift or carry them by the legs. Ducks usually lay early in the morning, but are inclined to drop their eggs anywhere, so it is best to keep them shut up until ten o'clock. Young ducklings should be kept out of the direct rays of the sun.

Whether turned off young or when mature, ducks will yield a good profit if rightly managed, and the number raised need be limited only by the capacity of the premises and of the man; the latter has much more than the former to do with success in duck raising. No one should go into duck farming unless he grows into it from small beginnings. It is doubtful if there is any one in this country who raises more ducklings than Mr. Jas. Rankin, and he began in a very small way years ago with a few common ducks. Now he has many incubators, brooders and brooder-houses doing all of the work possible in the business. The eggs are furnished by high-bred Pekin ducks that are given good care and close attention.

RAISING GEESE.

Although this country is better adapted (on account of the almost unlimited range obtainable) for raising geese than the old country, flocks of them are a much rarer sight than across the water. In many sections there is only an occasional trio or so kept by a thrifty German, who has not forgotton the comforts of a feather bed of his own raising.

Geese are great foragers and must have plenty of green stuff; they do better where there is plenty of range with convenient water; swamps and marshes are just the thing; but upland pastures and abandoned hilly farms do very well if water is provided. Many a worn-out piece of ground would yield a good profit if used as a goose pasture. The low lands along coasts and rivers may also be turned to very good advantage to their owners.

If, in starting out, geese are purchased, it is on the whole best to buy well mated old ones. The gander does not fully mature until the third year, and the goslings from younger parents will not likely be as strong or grow as large as those from fully matured birds. But there is nevertheless a redeeming feature in using young birds when extra early goslings are an object, as yearling geese will begin to lay earlier than older ones. Geese naturally pair, but two or three females may be kept with one gander, though he is very apt to have a favorite mate and the eggs of the others may prove largely infertile. Yet if the favorite begins to sit first, he will make love to the others and then there will be no trouble the next season; for this reason it is usually best to keep the same breeding stock for several years.

If the geese are given plenty of vegetables during the winter and not kept too fat, they ougbt to begin to lay in February or March and produce fifteen to twenty eggs before wanting to sit; often thirty eggs are laid if the birds are previously rightly cared for. When ready to begin laying, the old goose will usually carry bits of straws or stubble around to make a nest with; she should be shut up in a roomy kennel or shed-roofed box until she has laid and there she will usually return to deposit the rest of her litter.

HATCHING THE EGGS.

The first eggs should be set under hens, giving four to six eggs apiece; many prefer to have hens do all the incubating, as the goose is rather apt to crush the newly hatched gosling. The eggs may be hatched in incubators. If either by this mode or by hens, they should be well sprinkled the last two weeks of incubation, as the eggs are even tougher than those of ducks, and the young goslings will often have to be helped out. The time of incubation is twenty-eight to thirty days.

If the goose is to be the mother she should be left alone and the other geese kept from her, as she will probably resent interference in an unpleasant if not disastrous manner, and, especially if the gander is near at hand, as is often the case, the intruder may be roughly handled. Food and water should be kept near at hand, for if obliged to seek her food the eggs will very likely be chilled before she returns. When hatched remove the mother and brood to a large coop with a pen around it and plenty of shade, as the hot sun is fatal to young goslings. If a hen is set a day or so before the goose, all the goslings may be given the goose to care for.

Do not feed anything for twenty-four hours, then give same food as ducks for a few days.

The young goslings should be kept from the water for at least two weeks, after which they may run at large with their mother unless the weather is cold or rainy, when they should be kept under shelter. As soon as they begin to forage a feed of corn meal cooked with vegetables should be given morning and evening for several months; and they should be shut up at night. After this feed only whole corn at night if it is desired to raise them at little expense. They grow rapidly and are soon out of danger of chilling and are very free from disease.

If desired they may be sold as "green geese" when from six to eight weeks old; in this case they should be crowded from the start. Unless designed for breeding stock or kept for their feathers, they should be sold before a year old, as "old" goose is tough and brings a low price.

FATTENING FOR MARKET.

When fattening, feed three times a day on cooked corn meal and potatoes. (To be at their best they should have been fed at least twice a day from the very start). Geese are very fast friends, and when only part are being fattened the others should be kept out of sight or they will very likely pine for them and lose flesh. When marketing kill all of one lot at the same time, or the result just mentioned will follow.

Goose feathers always command a high price and should be plucked twice or thrice a year, according to climate. A good goose will yield a pound or more of feathers annually. In picking, follow the directions given for picking ducks.

Even though you have no suitable place to rear geese, but wish to enjoy the luxury of a home-grown bed or the royal repast of Christmas goose, a few may be profitably raised by herding them on the street, like cattle, and giving water for drinking only.

Many may have neglected to engage in this remunerative industry because of the common belief that geese are such "awful" eaters, but since, as has been mentioned, they live principally on grass during the summer, the cost of their feed is not great, considering the profit to be gained.

CHAPTER VI.

CAPONS AND CAPONIZING.

Capons are castrated cockerels, and when fattened are called the "finest chicken meat in the world." When caponizing, as the operation is termed, was first, practiced is unknown; it was in vogue as long ago as Shakespeare's time and has been very commonly and popularly carried on in France for many years. In some sections of a few Eastern States capons have been quite numerous for a few years, and the demand for them has increased their production, till now they are found quite frequently in the leading poultry markets of the East, and for two seasons at least, have not infrequently been seen upon the Chicago market.

A capon is neither hen nor rooster—he is a capon, and the reason that so much has been said about him in poultry and agricultural papers of late is that the makers of caponizing instruments have taken advantage of the fact that he sells for more per pound than either in the market, and usually weighs more at a given age than a rooster, to push their goods by showing the vast profit between having a lot of old roosters around and a lot of fine capons to turn off that have eaten no more than the cheap "drug on the market" rooster. In doing this they have done good missionary work towards reducing the number of cockerels that are annually allowed to grow from nice broilers or spring chickens to old roosters, and have spread broadcast much valuable information and furnished full directions for caponizing.

A set of caponizing tools usually consists of the following instruments, put up in a neat case, as shown in the illustration, with full directions inclosed:

FIGURE 1—Cords for quickly securing fowls to the table, rendering them unable to struggle in the least.

FIGURE 2—Knife or Lancet for making incisions.

FIGURE 3—A Spring Spreader for holding ribs apart after incision is made, and so constructed as to automatically suit any size fowl.

FIGURE 4—Sharp Hook to pick open the film-like skin after incision is made, with knife. This skin must always be picked open before you can proceed with the operation.

FIGURE 5—Probe for pushing back intestines when they crowd testicles, aiding the learner to slip loop over testicle

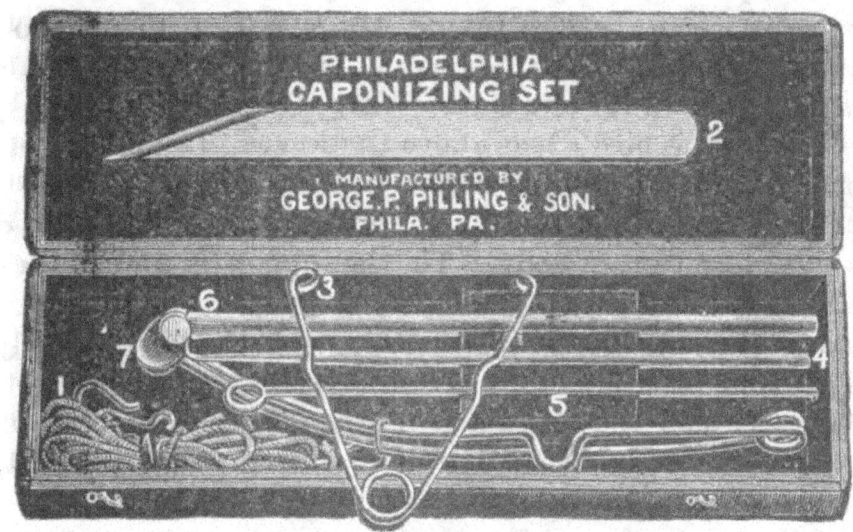

easily, and is also used to discover any foreign substances that may find their way into the cut before the operation is finished.

FIGURE 6—Caponizing Canula.

FIGURE 7—Curved Spoon Forceps, for removing any foreign particles that may remain, such as feathers, etc. These are used in conjunction with probe (Fig. 5).

The Directions for Caponizing, as given by Geo. P. Pilling & Son, Philadelphia, which are essentially the same as given by all manufacturers and may be relied upon, are as follows:

From twenty-four to thirty hours before performing the operation, select such cockerels as you intend to caponize (these should be from two to three months old), confining them in a clean and airy coop or room without either food or water. The best time to confine them is at early morning, as their long fast will then end about noon of the following day, at which time the operation is best performed. Should the day be cloudy or wet do not caponize them, but let the operation go until you have a bright and fair day. It is necessary that you have all light possible in the matter. If it be a cloudy day, and you decide not to caponize, the birds may be given a little water or food if necessary, but it is much better to avoid this, if possible, as it is very desirable to have their intestines quite empty, thus allowing their testicles to be more readily seen, besides giving the operator much more room in which to perform his work. Lay the bird on the operating table on its left side. Wrap the cord (Figure 1), twice around the bird's legs above the knees. In making one wrap only there is danger of the birds kicking themselves out of the loop. Hook the other cord once around both of his wings close to the body. To the opposite end of these cords attach a half brick or some equal weight, letting them hang over the sides of the table. This holds the bird securely. Have all your instruments in readiness that you may work quickly. Thread the Canula (Figure 5) with a strong and long horse-hair or fine steel wire (we think wire the best), letting the wire form a loop at the curved end and well out at the other end. Now after slightly wetting the spot proceed to pluck the feathers from the upper part of the last two ribs and just in front of the thigh joint. Pull the flesh on the side down toward the hip, and when the operation is finished the cut between the ribs will be entirely closed by the skin going back to its place. While holding the flesh back with the left hand, with the right hand take the knife (Figure 2) and insert it (cutting-edge away from you) between the last two ribs, cutting first down and then up a little ways, following the direction of the ribs, making the cut not over one inch long. Cut deep enough to go through the skin and ribs, being very careful not to go so deep as to cut intestines. There is little danger of doing this, however, if they are empty, as they will be from the bird's long fast. The danger of cutting the intestines is when they are full, as in this state they press against the ribs. Should the cut bleed, stop a moment, let the blood clot on the thin skin covering the bowels, and then remove it with Curved Spoon Forceps (Figure 7).

Now take the Improved Spring Spreader (Figure 3), press it between the thumb and finger until the ends come together, inserting the ends in the incision with the spring end toward the bird's feet (see operating table). Upon looking into the cut a thin tissue-like skin will be seen just under the ribs and enclosing the bowels. Take the sharp hook (Figure 4) and pick the tissue open so that you may get into the bird with the instruments. The breaking of this skin does not cause the least pain to the bird. One of the testicles will now be brought plainly to view lying close up to the back of the fowl. Sometimes both testicles are in sight, but this is not generally the case, as the other one lies beyond and more on the other side of the bird, the intestines preventing it from being seen from this opening. The testicle brought to view is enveloped in a film. This should be brought away with the testicle. Some people in caponizing tear the skin open and then take the testicle out. The danger in so doing is that if this skin is left there is danger of causing a "slip."

Now comes the only dangerous part of the whole operation—getting hold of and removing the testicles. But with a steady hand and plenty of light not one bird in a hundred should be lost. Attached to the testicle and lying back of it is one of the principal arteries of the fowl, and if this is ruptured it is sure to cause death. It is here that the improved Canula (Figure 5) proves of the greatest advantage. The hair (or wire) being small and very fine is easily slipped between the testicle and artery without injury to either, a clear, clean cut made that no other instrument can do. Take the Canula in the right hand and adjust the hair (or wire) in it so that a loop about one-half inch long will extend from small end of tube, leaving the two ends of the wire extending far enough out of the open end to secure a good hold. Insert the end of tube that has the loop on it very carefully and slip the loop over both ends of the testicle and entirely around it, hold end of tube close down to the testicle. When the testicle is entirely encircled by the loop, take both ends of wire (or horse hair) which comes out of the other end of tube with thumb and first finger, holding it tight, and draw up on it carefully but firmly, being particularly careful to have the loop around testicle. Keep end of tube very close to testicle all the time. If drawing up on the wire does not at once cut testicle, slightly turn from one side to the other (but not entirely around), then the testicle will come off. After removing it, carefully examine inside of bird to see that no piece is left in, and also to see that no foreign substances, such as feathers, etc., have gotten in. If any have, it is necessary to remove them, for, if allowed to remain they are liable to cause inflammation. Sometimes a feather or part of the testicle may drop among the bowels; if this occurs, move bowels around with probe (Figure 6) until the object is found, then remove with Curved Spoon Forceps. When the operation is performed remove the Spreader at once and the skin will very soon slip back over the cut and heal in a short time. Never sew the cut, as it will heal just the same as any other small flesh wound.

The bird can now be turned over on its right side, cut made and testicle removed in exactly the same manner as just described for the left side.

Both testicles may be taken out with the one incision but to the learner we would say this is attended with more difficulty than the two incisions. The other testicle being situated so far over on the other side, there is more difficulty in reaching it, besides danger in piercing artery running back of first testicle. To an experienced party there is no danger in removing both testicles from one incision, but to those who have not that degree of confidence given by practice we would recommend the two cuts. The bird recovers just as quickly as though one cut was made, and the operation is performed equally as quick, if not quicker. If both testicles are removed from one cut, the lower must always be taken out first, for if top one is first removed, the small amount of blood that may follow will cover lower one, keeping it from view.

As soon as the operation is performed the bird should be released. Carry him carefully by the wings, taking hold close to the body, to a cool coop or room without roosts or fixtures to induce flying, as such exercise is apt to open the wound and retard healing. Give him at once plenty of fresh water and all the soft food he wants. He will go to eating and take care of himself and show no signs that he has been in any way molested. The second or third day after caponizing look the birds over, as in breathing, the air generally gets under the skin, causing "wind-puff," a slight swelling. Simply prick through the skin at the sides with a sharp needle. Press the wind out gently and the capon will be relieved. This should be done from time to time for ten days or so when the wound will be so healed that it would be difficult to tell where the incision was made. At this time the capons may be given full range and should be fed as you would feed any poultry that is. growing rapidly. Capons do not mature till a year old and will often continue to grow till eighteen months or two years old, but usually it is not profitable to keep them till that age. Neither is it profitable to kill before a year old. The price and the time of year when the birds are a year old will help to determine whether to kill at a year old or keep a while longer. They are sometimes kept till older to mother little chicks.

THE OPERATING TABLE.

There are numerous styles of tables on which caponizing may be done. Some of them are quite elaborate, and have considerable machinery about them. For all practical purposes, the top of an ordinary barrel (see illustration A) meets all requirements, admits of the bird being easily secured, and brings it to the proper height for the operation. It costs practically nothing, as an empty barrel may always be had for a few cents if there should not chance to be one on the place.

The illustration B shows how a good caponizing board (or table) can be constructed by the use of Pilling's Improved Staple A to slide over the wings of the bird. The staple has two fenders, about one inch from the points, to prevent forcing the bird's wings too close together, as would be the case without them. The cross-bar on staple allows you to use the upper part for a handle. This will be found very convenient during caponizing. One point of staple is longer than the other; this enables it to enter the board much easier. By cutting six or seven holes in the board it will take any size bird. B is the strap loop with a pin

TABLE A—The above, photographed from life, illustrates method of holding fowl ready for caponizing.

across the top to prevent strap from falling through the board when not in use. C is the weight (at other end of strap) for keeping feet down.

This table is very good for those who propose caponizing on a large scale. The entire construction (as shown in illus-

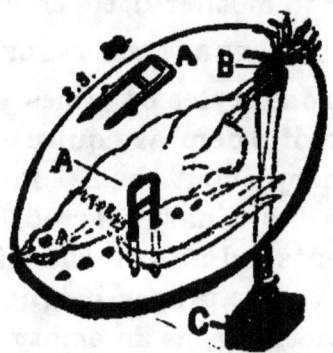

TABLE B—Can easily and cheaply be made and is suitable for those who intend caponizing on a large scale.

tration) is very simple and easily made. For those, however, who do not propose caponizing on an extensive scale, we would recommend the top of a barrel. Whichever method is used, make it a point to have plenty of sunlight and the table so situated that the light will strike squarely on the fowl. You cannot have too much light during the operation.

DRESSING CAPONS.

When your capons are ready for market select and shut up those you wish to dress and give them no food or water

for twenty-four hours previous to killing; then proceed to dress, as described on page 94. When completed, your bird will appear as above.

THE PROFITS OF CAPONIZING.

Samuel Cushman, Poultry Manager at the Rhode Island Agricultural Experiment Station, has been carrying on a series of experiments to get at the facts as regards the profits in caponizing. During 1891 and 1892 he experimented with five different lots. In each case cockerels of the same breed, size and age were fed in comparison with those caponized. From his "summary" we gather the following facts:

Caponizing was easily learned and successfully performed by following book directions, but more quickly and satisfactorily by witnessing the operation.

Birds apparently suffered but little pain from the operation and the per cent of loss was small.

Birds thus changed grew larger in frame, matured later, became quiet and contented, did not crow or fight and their flesh remained soft and tender.

Those weighing two pounds or less were most easily and safely caponized, but the larger the birds, provided they had not commenced to crow and their combs had not developed, the more quickly they recovered.

The only birds that died under the operation were those that had developed combs.

The old Chinese tools, when their use was understood, were found most satisfactory of all.

Of the Brahma-Cochin cross, it was seven months before the capons equaled the uncastrated birds in weight, and they did not average one pound heavier in ten months.

The Langshan rooster, although weighing but one-sixth of a pound more than the Langshan capon at the commencement of the experiment, kept ahead in weight for seven months.

The Plymouth Rock capon equaled the roosters in weight in less than two months and gained on them the rest of the season, but did not average more than three-quarters of a pound heavier at any time.

The Indian Game capons were five months in catching up with the roosters and were not a quarter of a pound heavier eight months after the operation.

The Brahma-Cochins gained the least during the first year, but made the largest and heaviest birds at eighteen months.

The Plymouth Rocks recovered less readily, but they

were operated upon when the weather was warmer, fifteen days later than the Langshan.

The Langshan was less affected by the operation, but was larger at the time it was performed.

Indian Games and their crosses were harder to do and should be taken when younger.

These experiments show less gain in weight as the result of caponizing than we were led to expect by published accounts. The tender flesh and the ability to take on fat seemed to be the only gain of importance.

The plan of sponging the wound immediately after the operation with an antiseptic solution, requires further study to get definite results.

By the aid of a physician's head mirror, we were able to operate quite satisfactorily by lamplight.

Those wishing to produce only a limited number of capons find it more profitable to secure the services of an expert, if one can be found within a reasonable distance, than to buy instruments and attempt the work themselves.

NOTES ON EXPERIMENTS.

Writers on caponizing compare the price of capons at maturity with the price of roosters at maturity. To be sure, if the cockerels are to be kept until they must be sold as old fowl, meanwhile fighting and running their flesh off, it would certainly pay well to caponize and keep them until nearly a year old; but they ignore the fact that early cockerels weighing five to six pounds when they are soft and tender, will bring as much or more per pound than a nine or twelve pound capon that has been kept twice as long. Quick returns are desirable, and the danger of loss by disease deserves consideration.

Had the Brahma-Cochin roosters of the second experiment been killed December 1, when they would have been beyond the tender stage, they would have dressed about seven pounds each and brought but twelve cents per pound (probably several cents less), or eighty-four cents. In July the same birds would have dressed about three and three-quarter pounds and would have brought thirty-two cents per pound or $1.20. If sold a month later, in August, although they would have weighed more they would have brought less, say four and three-quarter pounds dressed at twenty-four cents, or $1.14. This may be compared with a seven-pound dressed capon at twenty-two cents in January, $1.65 (forty-five cents for five months' feed) or a nine-pound dressed capon at twenty-six cents in April, $2.34 ($1.20 for bird in August and $1.20 for nine months' feed and care). This estimate is based on the weights of birds as recorded, and the market prices, in Boston, Mass., and New York City. These prices are for the very best quality of native fresh dressed poultry.

We found that capons (Western) were a drug in the Boston market in February, 1892, and could be bought for twenty-two to twenty-five cents per pound. The demand is less here than in New York, where they are more appreciated or better known.

The sale of capons has been hurt in some instances by selling pullets and roosters with small combs plucked like capons, but no one who knows need be deceived if the feathers and head are left on. When the market is bare of fresh summer raised poultry and everything but frozen stock or small-sized winter chicks are scarce (from January to June) there is a demand for a large tender "roaster" and capons fill the bill. They usually bring the best price in April and May when they are becoming scarce. Before chickens were so extensively raised at all times of the year by artificial means, capons brought high prices.

While visiting the New York markets we learned that great quantities of capons are received about January 1. The finest specimens and the greatest number are from New Jersey. None are received from the East, and those from the west are of poorer quality and contain a large proportion of "slips," although they are growing better each season.

At this time (December, 1892) there is hardly a limit to the demand for capons weighing eight pounds or over, and "Philadelphia" capons bring twenty cents and Western eighteen cents.

Large birds sell the best. The heavier the better. When ten-pound birds bring twenty-two cents, twenty-five cents will be given for twelve-pounders, and twenty-eight cents for those weighing fourteen pounds. Capons killed at ten or eleven months of age are preferred, as they get coarse and "soggy" if kept until twelve months old or longer. March hatched capons should be killed in January. The birds bought in January are placed in freezers and gradually sold during the winter. The supply is always exhausted before July. Prices begin to rise the latter part of February and continue to go up until there are none in the market. They are usually scarce in April and May. In the latter part of May, 1891, one New York firm could have sold a ton of capons at twenty-six to twenty-eight cents per pound in one day if they had had them. On July 7, 1892, capons weighing nine and ten pounds were selling at retail in Fulton market for thirty-five cents per pound. Seasons when grain is high, capons are not so extensively produced and the price is firmer.

Frozen capons cannot compare with those freshly killed in spring and early summer.

WHEN TO MAKE CAPONS AND THE TIME TO SELL.

Judging from these results and a study of the markets the best chance of profit by the production of capons would be in caponizing late chicks that ordinarily would be fit for market as broilers or roasters when the prices are the lowest, and too old to sell as tender chickens in January and February. Cockerels that were hatched in June, July or August, especially if of the large, early maturing kind like Plymouth Rocks and Wyandottes crossed on Brahmas or Langshans, castrated in September, October and November, and marketed in March, April, May and June, when they would have reached their best, would be the most profitable and bring the highest price. Such birds are often sold alive by the pound very low in the city markets or by those who have no room to winter them. Farmers who have cheap food, who are far from shipping points, and therefore kill and ship all at one time in cold weather, might profitably make capons of all roosters. Those who keep birds until maturity for their own table should do the same. There will be little gained by caponizing birds in May or June if they are to be marketed by Christmas, as the birds have not sufficient time to fill out.

James Rankin, the noted duck raiser, says: "I had some experience in caponizing twenty years ago, and found the business very profitable, as the birds readily brought from thirty to thirty-five cents a pound. At present, when capons bring but little more than half that price, it is not nearly so profitable, and for this reason I have given it up. I can get out a Brahma chick the latter part of January or the first of February, put him upon the market the first of June, when a little over four months old, when he will dress six pounds, and get $3 for him as a roaster, while a capon which I had caponized and kept for nearly a year, though he weighed eight or nine pounds, would bring me no more money and would have cost nearly double in both care and food. So I find decidedly more money in growing roasters than I can possibly find in caponizing."

Early this month, January, 1894, W. P. Leggett, of New York, says: Having looked carefully into the capon business, it suits me best to sell my surplus cockerels (100 to 250 per year) as they are. Turkeys and the Asiatic breeds of poultry take the edge off the capon market. Capons need fourteen to eighteen months to get full growth. Cockerels, if separated from all females early, and kept apart, get fat and develop breast very much more rapidly than those of same age running with the females, and do not get at all strong, and the sinews do not develop or get at all tough; therefore the meat is finer flavored and more tender.

Early-hatched Plymouth Rock and Light Brahma cockerels should now weigh eight to nine and nine to ten pounds respectively; capons of same hatch, very little if any more. For example, last year I had about 100 cockerels that I held till after the New York poultry show to sell as breeders. March 31, those not sold, I killed and sold at wholesale in Poughkeepsie, receiving eighteen cents per pound dressed. Light Brahmas averaged ten and one-half pounds, and White and Barred Plymouth Rocks nine pounds. Capons in same market were retailing at twenty-five cents, and only weighed seven to eight pounds. These cockerels were sold by their side for twenty-four cents.

Bear in mind the market has all to do with the profits.

If you are near a city, and can sell at retail or on order, you may come out ahead. For if you are going near your market and can deliver a few without loss of time or expense, do not compare your profits with one who has to make a trip expressly to reach his market. Let me here quote current prices from the finest retail market in Poughkeepsie: Chickens, good, eighteen cents per pound; extra, twenty cents; good capons, twenty cents; extra, twenty-two cents.

Never having raised cross-bred birds expressly for capons, I cannot speak of their weights, as I raise only pure-bred. It is asserted that first-cross cockerels grow faster and mature faster than thoroughbreds.

Another writer says: "The profit in caponizing seems to come from buying young roosters in the city market at live-weight figures and turning them into capons to feed through the winter. Few think it pays to raise capons from the eggs."

It seems to us that the whole question, like all others in the poultry line, resolves itself into one of circumstances, surroundings, locality and market facilities. At this writing, February, 1894, dressed roosters in the Chicago market are 5½ to 6 cents per pound and capons 12 to 12½ cents. These are not like the prices formerly received in Eastern markets, but they show how capons are appreciated. Money has been and can be made with capons. There are those who can make more with them than any other branch of poultry farming. The question for each to decide is, Is it I?

POULARDES.

Pullets are converted into poulardes by depriving them of the power of producing eggs. In France the ovary is usually extirpated. This is needless, as simply dividing the oviduct with a sharp knife is enough. The flank is exposed as in caponizing, but the incision should be made close to the side bone. The lower bowel will then be seen, and close beside it the oviduct, which is easily drawn forward with a blunt hook and cut across. This stops the development of the ovary, and causes the bird to attain more than its normal size.

Capons and poulardes, as a rule, will weigh about one-fifth more than the same birds in their natural state; but the flesh is whiter and more delicate, and the bird is plumper upon the table.   It seems to us that, however much cockerel meat may be be excelled by capon meat, it is hardly worth while to take the risk and trouble to poulardize pullets, for it is hard to beat a fat hen roasted, after her usefulness as a layer is past.

☞ The Philadelphia Caponizing Set, with full directions for using, can be had of the FARM, FIELD AND FIRESIDE, post-paid, for $2.50.

CHAPTER VII.

INCUBATORS AND BROODERS.

It is now generally admitted that chicks can be hatched and raised by incubators and brooders as well as by hens. Artificial incubation is no longer an experiment. Thousands and thousands of chickens and other poultry are profitably reared each year that never see a hen or other feathered mother.

The incubator way has many advantages over the old "setting hen." Those who tried to raise poultry before the days of incubators know how trying it was to have to wait till a hen wanted to sit before any early chickens could be had; sometimes it seemed as though biddie would never get ready for business, and when she did, if given a sitting of choice eggs the probabilities were that she would break several, and perhaps go and leave the rest just a few days before she should "come off." If none were broken it was very often the case that when off for feed the hen stayed from the nest so long that the eggs became chilled and the hatch was very unsatisfactory. All these troubles are gone since the days of incubators. If fertile eggs are put into a good incubator you may expect a large proportion of them to produce live chickens.

There are great factories both East and West where incubators and brooders are turned out yearly. Our Directory indicates where some of them are located. Each maker considers his the best, just as every mother thinks her baby is the nicest one, and will point out in what respect his excels all others. All incubators that are worthy the name are constructed with the same ends in view—uniformity of heat, sufficient moisture, and conditions as near like the ideal sitting hen as possible. The way heat is supplied, moisture obtained and the eggs cared for differs in the various

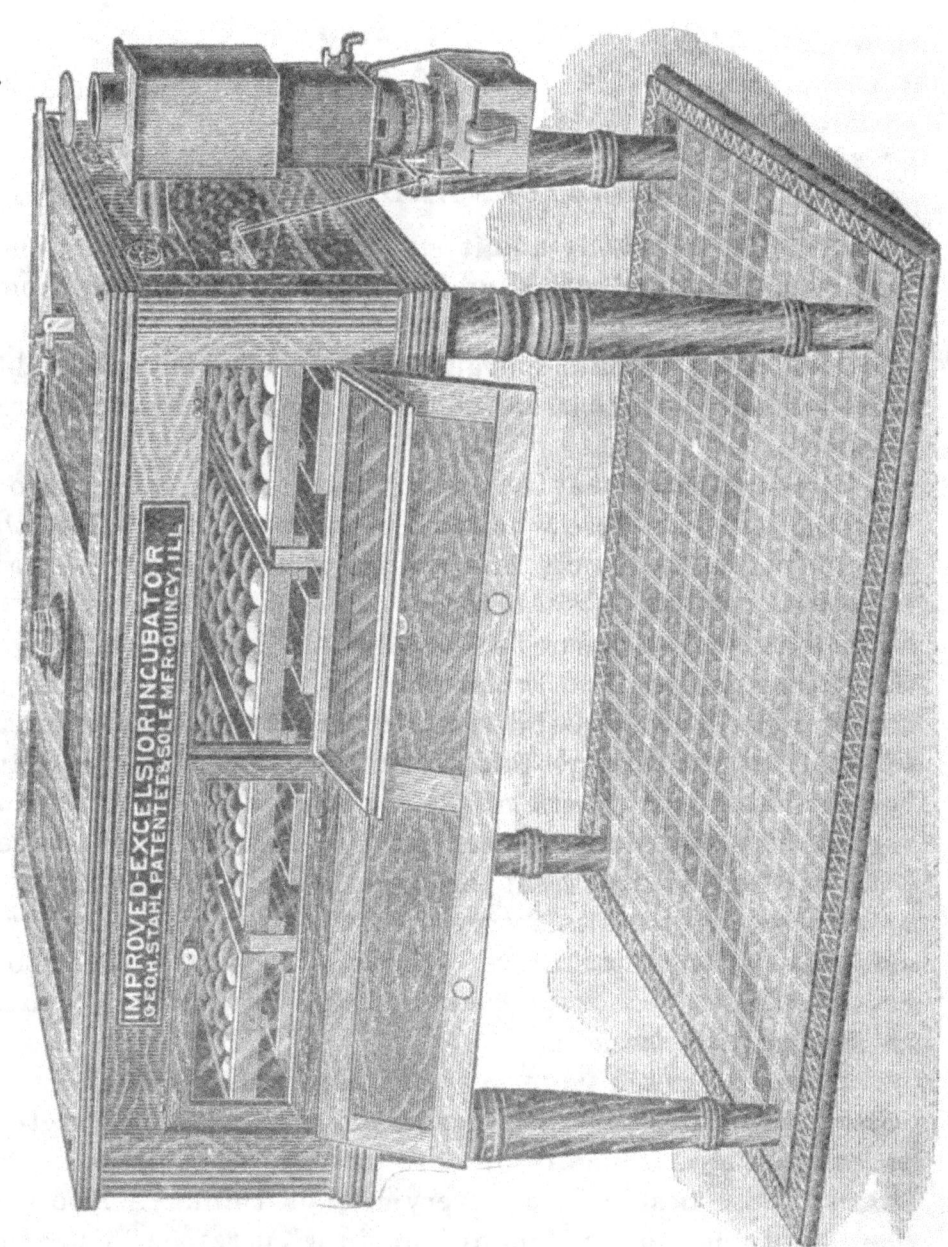

LATEST IMPROVED NO. 2 (200-EGG SIZE) EXCELSIOR INCUBATOR.

machines, and it would be impracticable, if not impossible, to give a minute account of them all in a book the size of this. Following is the detailed description of the Improved Excelsior Incubator, patented and manufactured by Geo. H. Stahl, Quincy, Ill. The accompanying illustration shows just how the No. 2 size, which holds 200 eggs, looks when filled and ready for business. As to its construction: The walls are double, even to the doors. The outer casing is made of thoroughly kiln-dried oak or ash, of first quality and paneled. The inner casing is made of pine. This gives a wall double all around, with a dead-air space between. This combination offers the greatest possible resistance to heat and cold, and its non-conducting power is so great that you can place the machine in a room where the temperature is sixty degrees, regulate it there, and the temperature may fall to zero or rise to ninety-five degrees, and the variation inside of the machine will be very slight. It has double doors, the inner one of glass, through which the eggs and thermometer may be seen. The outer one is of paneled wood, which is kept closed, except when it is desired to see the thermometer or turn the eggs.

MODE OF HATCHING.

As the hen supplies the necessary heat to bring forth her young by sitting on the eggs, they are heated from above. So with the Improved Excelsior Incubator—the heat is supplied from the top by the use of a tank which is filled with water, the water being heated by a lamp which is fastened to the end of the incubator, as shown in cut—the lamp chimney extending into the tank heater.

THE TANK.

The tank is built of the best grade of galvanized iron (heater lined with copper), and where exposed to heat or moisture is japanned, making it very durable. There are no pipes to clog up or get out of order, and the tank is easily removed by simply unscrewing the top of the incubator. The tank is so constructed that the water is kept in constant circulation through the heater, which keeps it hot, and imparts its heat to all parts of the egg-chamber alike.

The manufacturer claims this makes the most successful heating device ever used in an incubator.

THE TWO REGULATORS.

The regulator is to an incubator what the governor is to a steam engine; and as it would be folly to attempt to successfully run a steam engine without a governor it would be greater folly to attempt to hatch eggs in an incubator that was not furnished with a reliable regulator for upon the regulator depends the uniform and proper temperature and the proper ventilation necessary to hatch eggs.

On the Improved Excelsior there are two separate regulators, each acting independent of the other, so that either can be used separately or both at the same time—one regulating the flame of the lamp, the other acting upon a valve over the boiler flue, and in a measure controling the temperature of the water before it enters the tank.

The first is composed of a thermostatic bar, so placed and connected as to give many times the action and power of the ordinary rod. It is near the surface of the eggs, so as to regulate the heat at that level. This regulator is sensitive to the least change of heat, and very powerful; and instead of suddenly changing the flame from one extreme to the other—either very high or very low—it regulates the flame to give the required heat. The action is regular and graduated to the needs of the machine; if in a very warm r om, a low flame is produced; if the room grows colder the flame increases; and if the temperature of the room continues to fall, the flame grows larger until the full power of the lamp is turned on.

There is no clockwork, electric batteries or other contrivances to get out of order. Neither does it require an experienced person to operate it, as it takes care of itself.

The second, or valve regulator, is a single thermostatic bar, suspended above the eggs, in the egg-chamber.

The regulation of the machine is adjusted so that the lamp flame will keep the egg-chamber at the right temperature. The regulator of the escape-valve is then adjusted at 104 degrees (or at any desired point); should the temperature of the room from any unexpected cause rise above its usual degree enough to affect the incubator, or should the lamp, through neglect, fail to regulate perfectly, the valve will begin to open as soon as the heat of the egg-chamber reaches 104 degrees, thus making it impossible to overheat the eggs.

MOISTURE.

An incubator may not vary two degrees in temperature in twenty-four hours, the egg-chamber may be ever so well ventilated, and yet a large percentage of fertile eggs not hatch, simply because the eggs were not supplied with proper moisture. Ample provision is made for supplying moisture in the Improved Excelsior Incubator. Shallow galvanized iron pans filled with water are placed in the egg-chamber below the egg trays, and the heat slowly evaporates the water sufficiently to supply the necessary moisture to the eggs above.

THE EGG TRAYS

which are furnished with the Improved Excelsior Incubator are constructed of a material that, after much experimenting, has been found to be the best adapted to the purpose. The egg tray in the Excelsior is very simple. The eggs can be turned a trayful at a time. It is not done by turning a crank, but with an extra tray which accompanies each machine. This turns every egg without handling them. The trays are all made to fit precisely, so that in turning them the eggs are not jarred or broken. As shown in the illustration the egg trays are in one tier, thus insuring the same degree of heat to all eggs.

THE LAMP.

The lamp which supplies the necessary heat is an all metal safety lamp, constructed especially for the Improved Excelsior Incubator, and is supplied with a burner of large heating capacity.

The Cut-Off Burner is also constructed especially for this machine, and is very simple and reliable. It is made with a brass movable sleeve which encircles the lamp-wick. This sleeve is raised and lowered to regulate the flame of the lamp; this movement also keeps the burner free from cinders or charred wicking, and there is no possible chance for it to corrode, get sticky, and consequently become clogged up. In fact, it is a complete success.

The chimney furnished with the Improved 'Excelsior Incubator is also metal, with a mica opening, so that the blaze can be as easily seen and regulated as with a glass chimney. A reliable, especially designed incubator thermometer is furnished with each machine.

GENERAL DIRECTIONS FOR RUNNING INCUBATORS.

Every incubator and brooder manufacturer gives specific directions for running his machine and caring for the chicks after they are hatched and in the brooder, but there are some general directions that are applicable to all incubators:

1. An incubator can be more successfully operated in a room of even temperature, and it should be placed where no cold drafts can strike it.

2. Use fresh, perfect eggs, of even size and shape. No rough, ill shaped or overly large ones, or those of under size for the breed that produced them.

3. Keep the temperature of the machine at 103°. Do not cool the eggs.

4. Chickens may die in the shell from too much moisture, too high temperature, too low temperature, lack of constitutional vigor of parents, too frequent opening of the incubator, or because the eggs are from hens that are overfed and fat.

5. If the temperature is kept too low the chick may hatch, but it will be after time; if too high, the chicks may come out the nineteenth day.

6. Do not use eggs picked up anywhere they may be had and expect success.

7. Eggs from hens that are confined and overfed will not hatch, or will produce puny and weak chicks.

8. Incubators with a capacity of 200 eggs, or less, will give the best satisfaction to beginners—in fact, to anyone, unless situated so a large number of fresh, fertile eggs can be readily obtained.

9. The down of a newly hatched chick is no protection

from cold, and in winter great care must be exercised that no chick becomes chilled.

10. Like full-grown poultry, chicks need exercise. Keep them busily scratching in light litter furnished for the purpose.

11. Keep currents of air from passing over the chicks when in the brooders. If bowel disease appears it is usually due to colds induced principally from lack of warmth at night.

12. When chicks droop and appear sleepy, look for large gray lice on the heads or necks.

13. Dry feed is best for chicks, fed three times a day, but scatter millet seed or other small grain in the litter to induce them to scratch. A good authority on brooder raised chicks says they should have "rolled" dry oats for their first food, scattered where they can pick it up. Stale bread crumbs, dipped in fresh milk, are also good. These should be placed in little troughs. After the fourth day give the bread and milk for the morning meal, rolled oats at noon, and cracked wheat and cracked corn at night, with occasionally a little chopped eggs or meat. After they are ten days old feed them anything they will eat, compelling them to scratch as much as possible.

14. Supply water in such a way that the chicks cannot get themselves wet. Furnish grit in the shape of coarse sand, pounded shells, or some hard material.

15. The main requirement for successful raising of thrifty brooder chicks is warmth. If the chicks crowd together at night, you may be sure there is lack of warmth. If they separate under the brooder they are comfortable. In winter, the temperature of the brooder should be not less than 90° and not more than 100°. Examine the heating apparatus, as well as the position of the chicks, at bedtime, also early in the morning.

16. Keep the brooder clean.

17. Feed a variety of food, but let cracked wheat and cracked corn be a part of the ration after the chicks are old enough to eat them. Give cut clover hay for green food. Fresh milk may be given, but not sour.

BROODERS

are about as numerous as incubators, and it will answer our purpose to here describe only the Improved Excelsior Brooder, also made by Geo. H. Stahl, Quincy, Illinois. This artificial mother may be used indoors or outdoors—that is, may be used in the brooder houses in winter, or out of doors in warm weather. The accompanying illustration gives a good idea of this brooder, which, in its construction, is a happy combination of the "top" heat and "bottom" heat systems. The good points claimed for this brooder are:

1. An equable heat on all parts of the brood floor.
2. A perfect ventilation without draft on the chicks.
3. Entire freedom from chicks crowding.
4. Ease of cleaning—no small matter.
5. Freedom from all danger of fire—a point to be considered.

The construction is thus described:

The heater box has no bottom, but the top is a metal plate. Over this is an air chamber, and above this is a floor on which the chickens stand; and over this is the adjustable cover, surrounded by a woolen cloth, notched to allow the chickens to run in and out, as they would under the mother hen. Under this cover are four warm air pipes through which the warm, pure air comes from the air chamber above the metal plate. It rises to cover where it flows out, among and over the chickens, giving perfect ventilation, and, at the same time, carrying off all poisonous gases, and with the warmth of the floor keeping the excrement dry and odorless, and in condition to be easily removed with a brush. The cover being the same height as the chickens, they cannot climb upon each other, and there being no solid corners to crowd into, there can be no smothering. In fact, it is a perfect mother for chickens. It can be used anywhere.

If it is desired, the Improved Excelsior Brooder is furnished with a folding glass cover, as shown in the illustration opposite. This cover allows the brooder to be as light as possible inside, which some claim is a great benefit to the chicks.

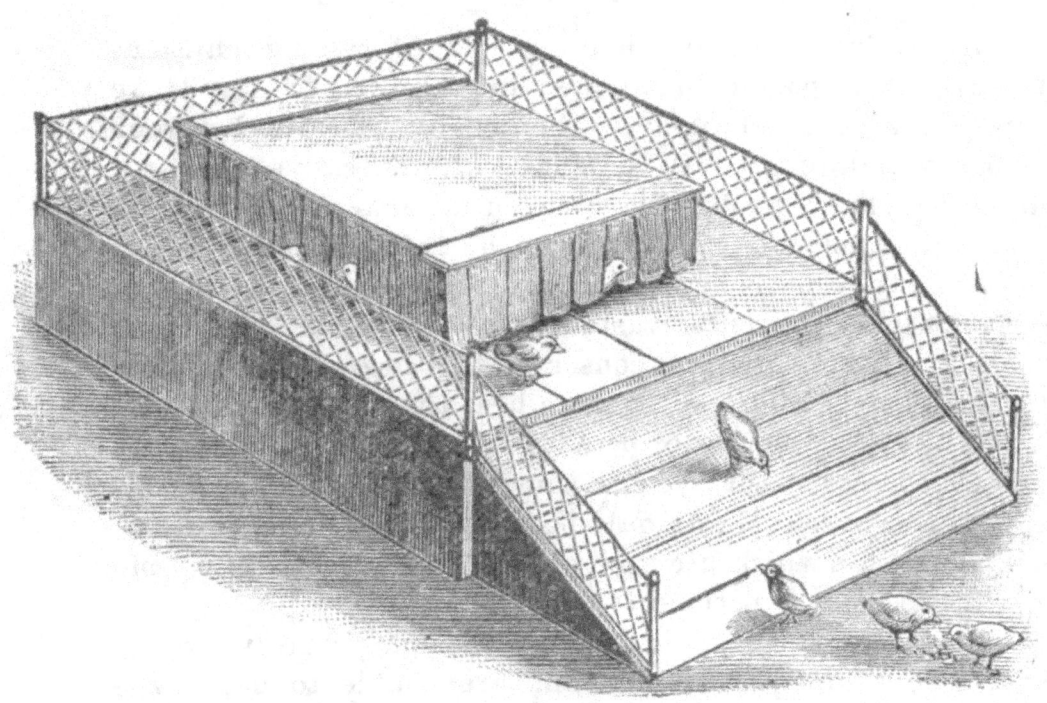

IMPROVED EXCELSIOR BROODER.

IMPROVED EXCELSIOR BROODER, WITH GLASS COVER.

TESTING EGGS.

In the days of the old hen method of raising chickens the eggs were put under the hen and allowed to stay there three weeks, if not broken in the meanwhile, whether good or bad or indifferent. In these days of progressive incubator hatching, the up to-the-time chicken raiser tests the eggs he places in his machine and removes the infertile ones, reserving them to feed his newly hatched chicks. This testing of eggs not only prevents waste of raw material, but leaves room for fertile eggs, and conserves the heat of the incubator, as a live chick produces more heat than a dead egg.

While the majority of persons who have good incubators make good hatches, there are some who would make decidedly better ones if they would post up a little on a few important points which are easily learned by practice of simple and inexpensive experiments.

Few persons understand testing eggs properly. Some have a very imperfect tester; some are unable to detect the infertile eggs closely—they cannot distinguish a dead germ from a live one, nor a weak from a strong one.

All eggs should be tested on the fifth or sixth day; at this test all clear or infertile eggs should be removed. To become expert in testing eggs during incubation, it is necessary to have a good tester. By use of a good egg-tester and the accompanying illustrations, any person can, with a little practice, learn to test eggs rapidly and accurately; the illustrations show exactly how the eggs look in the tester.

To become an adept in testing eggs for hatching, one has only to use a good tester, his eyes and a little judgment. As a lesson or experiment, try this on the fifth or sixth day after the eggs have been in the incubator. Break in separate saucers (carefully) one which you suppose (after examining with the tester and finding as shown in Fig. 1 on page 173) to be a good, strong, fertile egg; one which seems to be fertile, but weak; one that is doubtful—that is, one which you cannot decide whether it is fertile or infertile, and one that seems decidedly infertile. Break one at a time, ex-

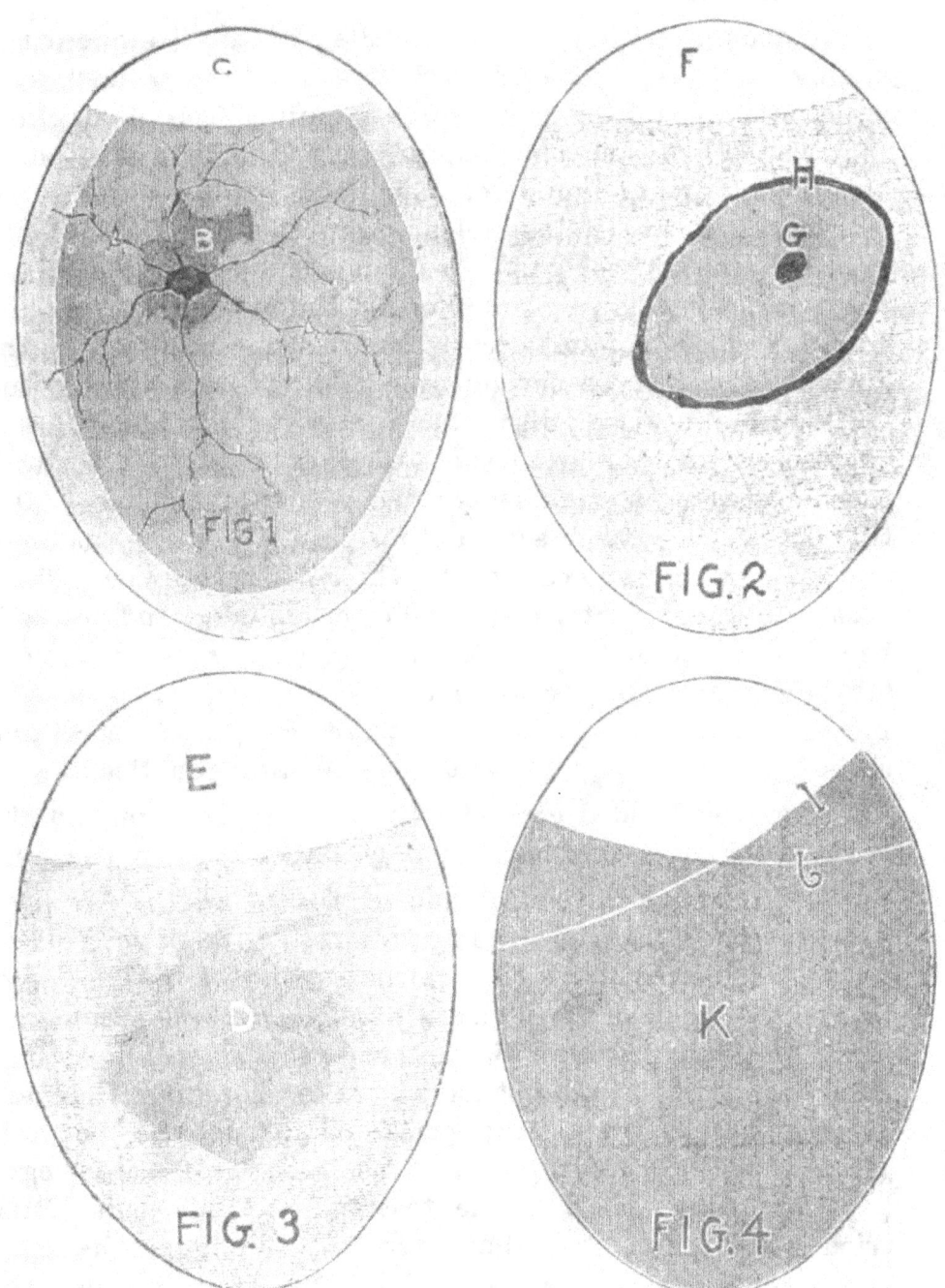

APPEARANCE OF EGGS AFTER BEING IN INCUBATOR.

amine carefully, making mental note. This first test should be on the fifth or sixth day.

A strong, fertile egg will, on the fifth day (temperature having been kept at 102°, 103° or 104°) show a dark spot which will float and show veins running from it, looking somewhat like a spider; a weaker one will show a spot but is cloudy looking and muddled. Such eggs are supposed to be fertile. Those which look clear are infertile. Do not mistake the yolk for the germ or chick. All infertile eggs are not perfectly clear. By breaking a few tested eggs and studying their contents, carrying in your mind's eye the appearance presented through the shell before breaking, you will be able to distinguish; having broken a strong, fertile egg, select another from the unbroken eggs, and see how it compares with the former. Then, having broken a fertile but weak egg, select another from the unbroken ones and see how well you can match the germ before you. Then break a few apparently clear and infertile ones, and you will be surprised to find some fertile eggs among them if your tester is inferior, or you are careless. You will also be surprised to find how easy it is to train the eye to detect and classify minute things by a little systematic practice.

There is decided economy in this egg-breaking at the commencement of business, for it will save eggs in the end.

Do not blame the sitting hen or the incubator for poor hatches unless you know that your eggs are fresh as well as fertile. Notwithstanding the possibility of fertile eggs bringing out chicks after being kept even three weeks, we never knew any one who did not prefer fresh eggs for setting. Some claim they would not have eggs for hatching that are over eight days old at any price—would not use them if given them. Others claim just as good results from eggs kept for weeks. Here is the testimony of one such: "We tried an experiment in that direction. We kept the eggs where they were as cool as possible, without freezing, the temperature not going below forty degrees nor above sixty degrees, and we turned them half around three times a week. Eggs that were kept in this manner for six weeks hatched as well as those that were fresh, and the chicks were strong and active."

If fresh eggs from healthy hens, fertilized by vigorous cocks, be used, they will hatch a large percentage of strong, healthy chickens.

In explanation of the illustration on page 173, Fig. 1 shows a strong fertile egg as seen in the tester on [the fifth or sixth day. B, the dark spot, is the live germ; AA are the blood vessels extending out from it. This germ, B, is seen by placing the egg against the aperture of the tester, and revolving it between the thumb and finger until the side on which the germ has formed comes nearest the eye. The spot B, will be seen plainly, often surrounded by a small cloud, as shown; the germ at this time is quite lively, and can be seen to move up and down. This is a strong, fertile egg, and should hatch under a good hen or in a good incubator. In a well fertilized egg the blood vessels should show plainly, but the germ is not always seen as plainly, varying with the color and thickness of the shell, and the power of the tester used. C shows about the average air bulb in an egg on the fifth or sixth day of incubation, though it may vary according to the freshness of the egg, and some eggs have larger air bulbs than others.

Fig. 2 shows a weak or imperfectly fertilized egg as seen in the tester on the fifth or sixth day. H is an oblong or circular blood vessel which has started, but nothing more; there is no heart, nor any part of a chick started. This egg will not hatch, but will decay if left in the hatcher, hence it should be discarded—removed from the incubator, or from under the hen. The small dark spot, G, is a weak germ, without blood vessels, only partially fertilized.* It has died after a start, and of course will not hatch. Both H and G

---

*"A partially fertilized germ" means one that from one of several causes was not strong enough to live and grow. Among these causes are cocks that are too old, an insufficient proportion of male birds for the number of females; old or debilitated hens, overfat hens, too close confinement of breeding stock.

You may find G (Fig. 2) among eggs which you believe or know are not over a week old, and ordinarily the eggs were good and fertile. It frequently happens that an egg will remain in the nest while several, or maybe a dozen hens lay there, and the succession of layers keep the egg warm enough to start incubation, or it

may sometimes be seen in the same egg. It will not hatch. F, the air space, may be seen in the same egg. The egg may be comparatively fresh, and yet show both H and G.

Fig. 3 shows a stale egg, a clouded egg, a doubtful egg. A stale egg is generally distinguished by the air space, E, being very large on the fifth or sixth day, as shown in Fig. 3, though all stale eggs do not show a very large air space, but when an egg does show it, it is pretty good proof the egg is stale. When an egg shows a clouded, muddled appearance, as indicated by D (which generally moves about when the egg is turned before the tester) it is certainly stale, and will not hatch. Do not confound the fresh egg which is not fertile with the stale egg; in an infertile fresh egg you can see the yolk, which will look somewhat darker than the rest of the egg, but does not look muddled.

Fig. 4 shows a live egg on the sixteenth day. K is the space occupied by the chick; the lines I and J show the air space, which may be on top or at the side, as indicated by the respective lines. This is about the average air space on the sixteenth day, but it will vary according to the thickness of the shell, and age of the egg when set; then, some eggs are not as full as others. At this stage of incubution (sixteenth day) a live chick darkens the egg, except the air space, when

---

may happen that some eggs may have been subjected to a heat of 100 degrees, in some warm place, unknown to or unnoticed by you. In either case, these eggs are taken from the nest or warm corner to a cooler place, and kept a few days, or over night, until a sufficient number has been accumulated to set; they become cold, and the germ dies before they are put under the hen or in an incubator.

In testing the first time, on the fifth or sixth day, a dead germ may be mistaken for a live weak germ, and if left in the incubator for three weeks would decay; so it is always best to test the eggs again on the tenth day, and remove all that have been marked doubtful and prove not good.

Perhaps some may think it is just as well to leave all of them in until hatching is finished, but this is not right; the decaying eggs generate objectionable gases, and if broken are very offensive. Besides, a dead egg or an infertile egg does not contain the animal heat that a live one does and is apt to have an undesirable effect upon the egg next to it, either under the hen or in the incubator.

An infertile egg—one which has not been impregnated, and in which life will never start or develop—is clear when shown at the tester. This egg, under the powerful lens of a first-class tester, will show the yolk, which must not be mistaken for a doubtful or a fertile egg. Be sure and use only a good tester.

seen with the tester, and, by watching the line I or J, the chick may often be seen to move.

Eggs should be tested in a warm room, one tray at a time.

The chick is harder to see after the seventh day, because the egg becomes more clouded by the growing chick.

HOW TO FEED INCUBATOR CHICKS

Miss H. M. Williams, of that broiler center, Hammonton, N. J., gives the following directions for caring for and feeding incubator chicks:

"The chicks, when taken from the incubator, should be placed in a brooder, in a warm room, with the thermometer of the brooder at 90°. Lukewarm water, and dry pin-head oatmeal, should be placed within their reach. They will feed as soon as they require it. The second day feed stale baker's bread, slightly moistened with hot water and dusted with black pepper. The third day feed a hoe-cake made of the following mixture: Three quarts feed meal, one quart wheat bran, one teacup ground meat, one teaspoonful baking soda, three tablespoons vinegar; wet the mass with cold water, or skim milk is better if you have it. Don't make it too wet, it must be a dry crumble. By dry crumble I mean so that it will break up in the hands, then crumble easily. Bake in a moderate oven two hours. When nearly cold crumble finely and feed. (The hard crusts may be soaked in milk and fed to old hens.)

"This feed may be continued for one week, in addition giving chopped cabbage, mashed potatoes and a small portion of boiled meat. (The proportion of cabbage, potatoes and meat being two tablespoonfuls after it is chopped to every hundred chicks.) Give fresh water every morning, and never allow the vessels to get empty. There should be enough vessels to admit of the chicks drinking without crowding and wetting each other as they will do in their eagerness to drink, especially in the early morning. At the end of ten days they should be removed to the brooder house and given the following feed four times a day: Five quarts corn, oats and wheat ground together, one quart wheat bran, one quart ground meat and one pint bone meal

—the whole mass scalded and allowed to stand one hour to swell. A small box of ground meat, charcoal and oyster shells should be constantly before them.

"We have discarded hard boiled eggs entirely, substituting raw eggs slightly beaten, into which crumble stale baker's bread enough to soak it all up. Four eggs to every hundred chicks may be given to advantage daily. Raw eggs seem to correct bowel trouble. Whole wheat may also be given with benefit. The chicks can be taught to eat it by mixing the whole grains with some that have passed through a coffee or bone mill, and been slightly crushed.

"Chicks cared for in this way can be sent from incubator to market in eight weeks weighing two pounds dressed."

A HOME-MADE BROODER.

Some who have no incubator wish a brooder, but prefer to make one rather than buy; so, also, those who have a small incubator and raise only a few chicks. To help such we give the following directions for making a home-made brooder, as furnished the Rural New Yorker by C. E. Chapman:

Make a box fifteen inches high and two feet square, for 100 chicks. Nail strips of tin on the upper edges and put on a sheet-iron cover. Cut an inch hole in the center of the sheet-iron, and put in an inch tin tube; let it fit tightly. It should be eighteen inches long. Make a frame two inches high and the size of the box and nail it on top of the sheet-iron. Bore some half-inch holes in it on one side so they will admit the air just above the sheet-iron. Make another box a foot high, the same width, and a foot longer than the first. Cut a hole in the floor and fit in a quart can, the bottom of which is out; punch an inch hole in the center of the top, and several quarter inch ones around it near the top. Place the second box on top of the first, and the tin tube will go up through the can. Put a piece of tin on the bottom of the second box, but keep it from touching the boards by nails partially driven in. This is to keep the boards from getting too hot, and it need not be over a foot square. Put a door in the front of both boxes. The upper box should have some glass in the sides and door; cut a hole in the back so that the chicks can go on to the feeding floor, and use a piece of cloth to keep in the warm air. The feeding floor is attached to the main box by hinges, and when down rests on the ground in a slanting position. Nail slats on it, so that the chicks can climb it. A hole cut in its side lets the chicks out when the floor is let down. A pin slipped into a hole in the end under the feeding floor keeps it from dropping down when you do not wish the chicks to go out. Put a slanting roof on top of the second box and a slanting addition, with doors for a cover, around the feeding floor. Put a pane of glass in both doors over the feeding floor, and a sheet-iron cover on the second box, before putting on the roof.

Between the sheet-iron on top of the second box and the roof is an air chamber. The small tin tube just comes up through the sheet-iron and heats this chamber. A small hole in the end of the brooder lets out the air and the fumes of the lamp. Now if a lamp is lighted and placed in the lower box the sheet-iron will become hot, and the fumes of the lamp ascend to the upper chamber and pass off without coming in contact with the chicks. (A tin lamp holding one or two quarts of oil is best. Have a large burner and sheet-iron chimney. A flat lamp that will not tip over is best; any tinner will make one cheap. Let the lamp be set close to the sheet-iron, leaving only enough space to prevent smoking.) The air above the sheet-iron will become hot, and pass up through the quart can into the second box above the chicks, near the top. Only a portion of the floor will be warmed from below, as the second box is longer than the first, and "bottom heat," which many think causes leg weakness, is avoided.

I made another stand which stood on legs, and fitted up to the brooder so that when the front door was open the chicks could go out into it. In this box there was wire netting over the top, and grit, water and fine seeds were in it. This was intended for an outdoor brooder and is better than a "chicken house," as fewer birds are kept together, and the air is purer and more uniform in temperature. A thermometer should be hung inside and the air should be kept at 95° to 100° the first week, and at not less than 90° at any time. If too warm, the chicks can go outside or out into the feeding room, but if too cold they will crowd and smother each other, and die from diarrhea. Let them be so warm that they will keep apart; use sand or earth on the bottom of the brooder; clean out often, and with vigorous chicks all should be raised.

AN INTERESTING EXPERIMENT.

The following, relating to an experiment by a French scientist, though of little practical use to the incubator manager, is interesting to any one who finds pleasure in understanding the development of infantile chicken life.

The shell on either side of an egg was removed, without injuring the membrane, in patches about the size of the diameter of a pea. In these openings bits of glass were snugly fitted. As to the results the experimenter says:

"I placed the egg with the glass bull's-eye in an incubator run by clockwork and revolving once each hour, so that I had the pleasure of looking through and watching the changes upon the inside at the end of each hour. No changes were noticeable until after the end of the twelfth hour, when some of the lineaments of the head and body of the chick made their appearance. The heart appeared to beat at the end of the twenty-fourth hour, and in forty-eight hours two vessels of blood were distinguished, the pulsations being quite visible. At the fiftieth hour an auricle of the heart appeared, much resembling a lace or noose folded down upon itself. At the end of seventy-two hours we distinguished wings and two bubbles for the brain, one for the bill and two others for the forepart and hindpart of the head. The liver appeared at the end of the fifth day. At the end of 131 hours the first voluntary motion was observed; at the end of 148 hours the lungs and stomach had become visible, and four hours later the intestines, the loins, and the upper mandible could be distinguished. The slimy matter of the brain began to take form and become more compact at the beginning of the seventh day. At the 190th hour the bill first opened and the flesh began to appear on the breast. At the 194th hour the sternum appeared. At the 210th hour the ribs had begun to put out from the back; the bill was quite visible, as was also the gall bladder. At the beginning of the 235th hour the bill had become green, and it was evident that the chick could have moved had it been taken from the shell. Four hours more and the feathers had commenced to shoot out, and the skull to become gristly. At the 264th hour the eyes appeared, and two hours later the ribs were perfect. At the 331st hour the spleen drew up to the stomach and the lungs to the chest. When the incubator had turned the egg 335 times the bill was frequently opening and closing as if the chick was gasping for breath. When 351 hours had elapsed we heard the first cry of the little imprisoned biped. From that time forward he grew rapidly, and came out a full-fledged chick at the proper time."

REMINDERS ABOUT INCUBATORS.

To test a thermometer place the bulb under the wing of a hen close to the body, shutting the wing upon it. The heat should be 104°. If the thermometer records 102° it is incorrect, but may be used, only the 102° should be taken as 104°. That is, allow 2° for its variation. The heat is the same under a hen when she first begins to sit as under one not sitting.

Do not be afraid to watch your incubator during the night when you expect 100 chicks to come out. For, though an incubator will regulate, it has not brains.

When the eggs are hatching ask the curious visitor to bring his visit to a close.

No matter how good the regulator, do not expect to have a good hatch without work.

Read as much as you have a mind to about running an incubator—experience will teach you more.

The incubator should be warm before the eggs are put into it.

Remember incubators are not toys to be played with by children. They are made for business, and for that business they should be managed with business methods, not child's play.

The best incubator is the one that best suits your wants and that you best understand. Each has its good points; the one that you can run best is the best for you.

If you expect to run an incubator to its full capacity, better have a brooder of twice the capacity of the incubator Or, what is better, have two brooders the combined capacity of which equals that of the incubator. That is, small brooders give better satisfaction, at least to the unprofessional, than large ones.

BROODER HOUSES.

Where the broiler business is extensively carried on the brooding of the chickens is done in houses devoted to that purpose. These are warmed by various systems, each owner selecting the system that best suits his fancy or the size of his pocketbook. The illustration on page 183 shows the exterior of a good one where glass is used in the roof. Some prefer to put the glass under the eaves. The second illustration on page 183 represents the interior of a brooder house and shows the arrangement of the brooders.

EGGS FOR HATCHING.

To have good eggs for hatching, either in an incubator or under a hen, we must first of all have good, healthy, vigorous stock to produce the eggs, and to all who contemplate buying eggs for hatching, I will say, be sure and find out just how the fowls are kept that lay the eggs, and what condition they are in.

One very vital point is to see that too many cocks are not kept; especially is this the case where all the fowls run together, if each cock has his own yard and hens it is not so important. Where too many cocks run with the same flock of hens the eggs are never good for hatching, for more deformed chicks will be hatched from such eggs than any other kind.

Exact rules can not be given, of course, but approximately, the following may be considered about the right mating: Leghorns, Spanish, Hamburg and Game, eighteen to twenty hens with one good cock, not less than fifteen months old for best results. Plymouth Rocks, ten hens with one cock. All the large, heavy breeds, including Brahmas, and all the Cochin family, five to eight hens with one cock.

To get a first-class hatch eggs must never be allowed to

BROODER HOUSE.

1. **Outside appearance, as seen from alleyway.**
2. **Upper doors and Brood-top removed.**
3. **Upper doors removed and Brood-top in position.**

INTERIOR OF A BROODER HOUSE.

get dirty; there is but little choice between a washed egg and a dirty one, and you are not sure of a first-class hatch with either kind, although the washed ones, if the washing is done with clear water, stand the best chance.

Some claim there is nothing as good for a nest as clean sharp sand, for two reasons; it soon cleans lice off of hens, and eggs can never get dirty on clean sand, nothing can come off the sand that will stop the pores of the shell.

Eggs for hatching should always be carefully gathered twice a day, and as fast as laid in cold weather; should be kept in a cool place where it is neither dry nor damp—that is, damp enough to mildew or mold anything. A good cellar where the temperature is from 40° to 50° is a good place; near 40° the best.

The proper position in which to place them while being held till wanted for setting, is a disputed point; some say large end up, some small end up, others on side and turn over daily; if they are not kept more than a couple of weeks it will make but little difference which end is up.

Before the eggs are put in the incubator or under the hen they should be carefully examined with a good egg tester. All that can be done at this test is to see if the eggs are fresh and perfect, nothing whatever can be told about the fertility, but all eggs that are not perfect both in shells and contents should be thrown out; any that have small specks of any kind visible inside, all with shells that appear cracked, all that have the yolk adhering to the shell, those that have thin, rough shells, especially at the large end. Use nothing but perfectly shaped, clean eggs if the best re sults are expected.

Incubator manufacturers have egg testers for sale, but if you do not wish to buy one, a very good one, for at least this preliminary examination, may be made as follows: Have a darkened room. In the curtain, blanket or whatever the room is darkened with have a hole through which light can come into the room. Have a cigar box with two egg-shaped holes opposite each other. Over one tack a piece of leather or felt in which is an egg-shaped hole just large enough to fit nicely around an egg when placed in it. Hold the egg in this and before the hole in the curtain, look through the opposite hole of the box and you can detect any shortcomings in the egg that you did not observe by daylight. The same work can be done at night if a good lamp is used in connection with the box.

HOME-MADE INCUBATOR.

For those who wish to try their hand at making an incubator at home we give the following directions for which we are indebted to P. H. Jacobs, former editor of the FARM,

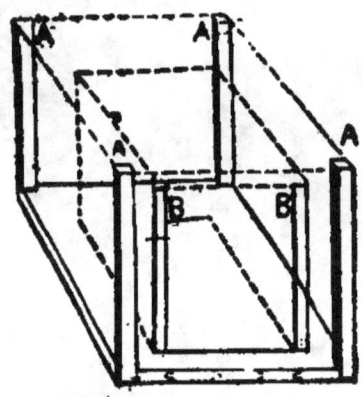

Fig. 1. Inner Box.

FIELD AND FIRESIDE and now editor of the Poultry Keeper. It is used at Hammonton, N. J., and the materials cost from $6 to $10, according to size. First, get good boards an inch thick and a foot wide. Cut them forty-six inches long for your floor, and have the floor forty-two inches wide. Place four posts, which are twenty-four inches high, at each corner (Fig. 1) marked A A A A, and two posts (B B) in front, the front posts to be eighteen inches high. Make posts of 2x3 strips and nail them securely to the floor. Fasten the floor boards together by strips underneath, using as many as desired. The four corner posts are for your

OUTER BOX.

This box when finished is four feet long and forty-four inches wide, outside, provided it is made of boards one inch thick. Including its top and floor, it is twenty-six inches high. Nail on your side boards. Let rear and front end boards cover ends of side boards. After the tank is in, and the top of the inner box is on, cover inner box with sawdust, and nail down the top of outer box. Tongued and grooved boards should be used for every part of the incubator except the floor, which should be of heavy boards. All the measurements given here, however, are for boards one inch thick, but three-quarter stuff may be used if desired.

THE INNER BOX.

This holds, or rather comprises, ventilator, egg-drawer, and tank. It is forty inches long and thirty-two inches

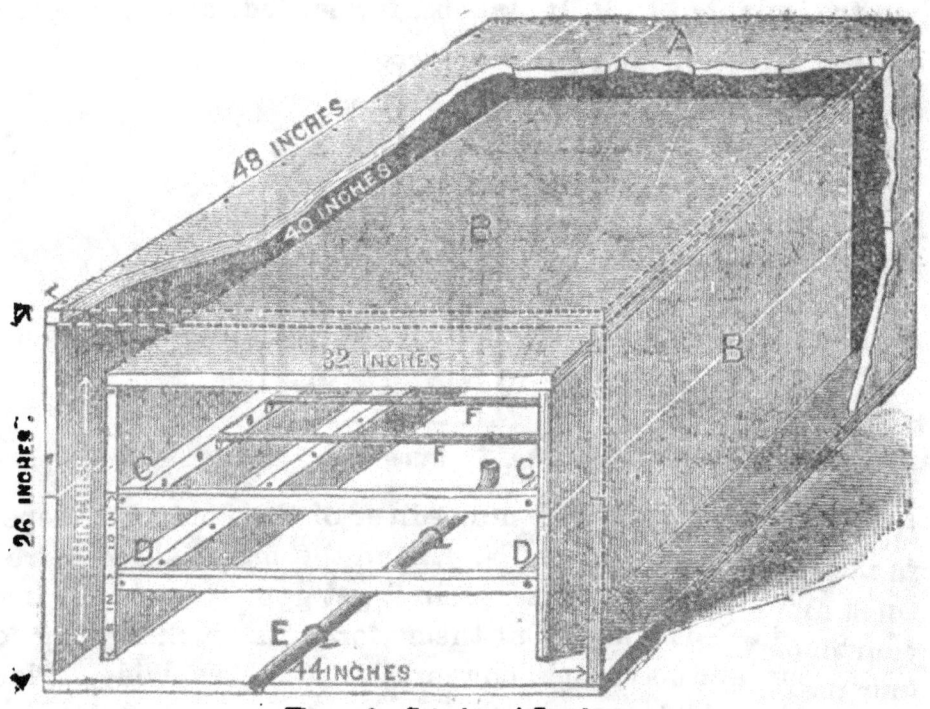

Figure 2. Interior of Incubator.

wide, outside measurement, and must hold a tank 30x36. The side boards are nailed to the posts B B (Fig. 1) and front boards of outer box, and fastened at the rear end by the rear boards being nailed to the ends of the side boards. Cleats are put on end and sides (on the floor), to fasten the inner box to the floor. Nail the bottoms of the side and rear end boards to the cleats.

To make the inner box, refer to Fig 2, which has portions of the outer and inner boxes torn away to show interior. A is the large or outer box; B is the inner box; C C are strips one inch wide and one inch thick, nailed to sides of inner box. D D are strips one inch wide and one inch thick nailed to sides of inner box. The strips C C, with iron rods half an inch thick (F F F F), hold and support the tank. Let ends of iron rods extend a little into sides of inner box, to assist in supporting the weight of the water. The strips D D are to hold the egg-drawer. E is a tin tube one and one-fourth inches in diameter and two feet long, placed in the front part of the

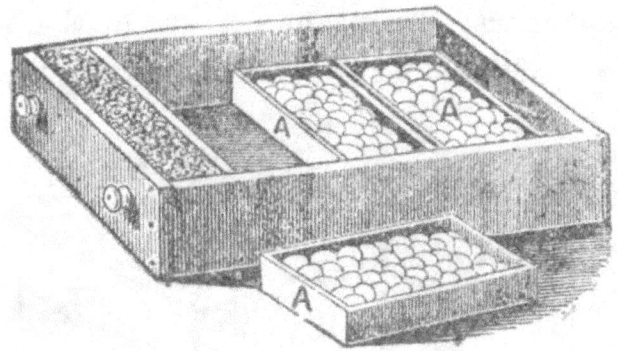

Fig. 3. Egg Drawer.

ventilator to admit air. Observe, however, that Fig. 2 does not show the sawdust in front, as will be explained.

We will now take up the separate parts. First is the

VENTILATOR.

This is simply the bottom of the inner box, being under the egg-drawer, five inches deep and thirty inches wide (the side boards of the inner box being its sides). The front end is boxed off, which includes the front boards and also the sawdust, thus making ventilator, inside measurement, thirty-six inches long. E is the tin tube, for the admission of air, before mentioned. Use no sawdust in the ventilator, but paper the bottom well and close, so as to have no air enter except through the tin tube. The tin tube is open at the front on outside of incubator and enters into ventilator.

EGG-DRAWER.

The egg-drawer goes under the tank, and rests on the strips D D (Fig. 2). The egg-drawer is four inches deep, outside measurement. It is thirty-nine inches long, outside measurement (which includes the boxed-off portion in front of drawer), and is thirty inches wide. Three movable trays, each one and one-half inches deep, are fitted in egg-drawer. Nail strips one inch wide and five-eighths of an inch thick, one inch apart, the length of the egg-drawer (but not under boxed-off portion) for the bottom. Mortise ends of strips in egg-drawer, so as to have the bottom smooth. Tack a piece of muslin on these strips (thin muslin is best) and tack it on the inside of the drawer. Now nail strips to bottoms of trays (use lath, if desired, cut to one inch width) but you need not mortise them. Simply nail them on the bottom, one inch apart, running lengthwise, and tack muslin on the bot-

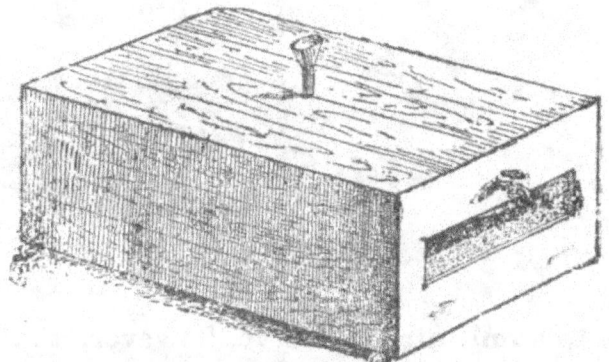

Fig. 5. Incubator ready for the Egg Drawer.

tom of the trays, inside, in the same way as for egg-drawer. The inside of the drawer will be three and three-eighths inches deep. The sawdust in front of egg-drawer (the boxed portion) fits in boxed front of incubator (see Fig. 5). Put a board cap on outside of egg-drawer, at front end, to exclude air.

THE TANK.

This is 30x36 inches and seven inches deep. It is supported by the strips C C, and rods F F F F (Fig. 2). Being thirty-six inches long, it goes close up to the back boards of the inner box, the front being inclosed by a sliding board, secured with upright strips at each end of board, one inch in diameter (so as to remove tank when necessary), which leaves a small space in front of the sliding board to be filled with sawdust. Have the tank tube in front only long enough to extend through the sawdust in front, and have your fau_ cet to screw into this tube, the tube being threaded. The tube on top of tank should be long enough to extend through

the tops of both boxes (outer and inner, through the sawdust), and should, therefore, be seven inches high from top
of tank, as is seen in Fig 4. When the incubator is ready,

Fig. 4 Tank.

.we have Fig. 5, which shows the sawdust packing in front,
by looking into the opening into which the egg-drawer enters when filled with eggs. Fig. 6 shows the incubator as
if cut in half lengthwise, and displays all the positions.

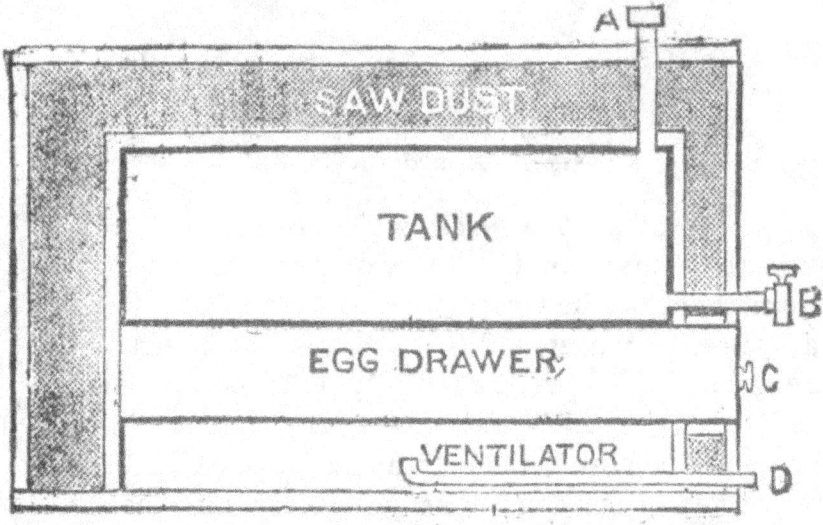

Fig. 6. Sectional View of Incubator.

What is meant by the "boxed-off" portion in front is that
portion filled with sawdust in front. The side boards of the
inner box are joined, on their front ends, to the front boards
of the outer box, being also nailed to the two short middle
posts. Fill in between the boxes with sawdust, and if sawdust is scarce, use chaff, oats, finely cut hay (rammed down),
or anything that will answer, but sawdust or chaff is best.
In Fig. 6, A is the tube on top, B the faucet in front, C the

opening for the egg-drawer, and D the tube to admit air into the ventilator. This tin tube should be as close to the bottom of the ventilator as possible. When making incubator do not forget to cut holes for tubes of tank and also for air tubes to come through, and then putty around them The tank should be made of galvanized iron.

DIRECTIONS FOR OPERATING.

Each tray holds about eighty eggs, laid in promiscuously, the same as in a nest, making total number for incubator

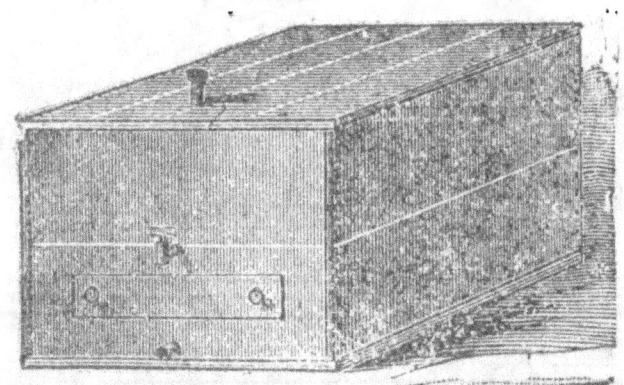

Fig. 7. Incubator Complete.

240 eggs. First fill the tank with boiling water, but never allow it to remain in the tube on top, as it thus increases pressure; hence, when tank is full to top of the tube, draw off a gallon of water. Fill it forty-eight hours before putting eggs in, and have heat up to 115° before they are put in. As the eggs will cool down the heat, do not open the drawer for six hours, when the heat should be 103°, and kept as near that degree as possible, until the end of the hatch. It is best to run it a few days without eggs, to learn it thoroughly.

Place incubator in a place where the temperature does not fall below 60°. As the heat will come up slowly, it will also cool off slowly. Should the heat be difficult to bring up, or the eggs be too cool, you can raise or lower the trays, using small strips under them. You can also stop up or open the air tube in the front opening of the ventilator whenever you desire. When the eggs are put in, the drawer will cool down some. All that is required then is to add about a bucket or so of water once or twice a day, in the morning and at night, but be careful about endeavoring to get up heat suddenly, as the heat does not rise for five hours after the additional bucket of water is added. The cool air comes from the ventilating pipe, passing through the muslin bottom of the egg-drawer to the eggs. Avoid opening the egg drawer frequently, as it allows too much escape of heat, and be careful not to open when chicks are hatching, unless compelled, as it causes loss of heat and moisture at a critical time. Cold drafts on the chicks at that time are fatal. Do not oblige visitors. Be sure your thermometer records correctly, as half the failures are due to incorrect thermometers, and not one in twenty is correct. Place the bulb of the thermometer even with the top of the eggs, that is, when the thermometer is lying down in the drawer, with the upper end slightly raised, so as to allow the mercury to rise, but the bulb and eggs should be of the same heat, as the figures record the heat in the bulb, and not in the tube. Turn the eggs twice a day at regular intervals—six o'clock in the morning and six o'clock at night. Do not let them cool lower than 70°. Turn them by taking a row of eggs from the end of the tray and placing them at the other end, turning the eggs by rolling them over with your hand. By removing only one row you can roll all the rest easily. Give no moisture the first week, very little the second, and plenty the third week. Do not sprinkle the eggs.

For moisture, put a wet sponge, the size of an egg (placed in a flat cup), in each tray, the second week, and two sponges in each tray the third week. Do not put in sponges until you are about to shut up the drawer, after turning. Wet the sponge by dipping in hot water. After the first ten days the animal heat of the chicks will partially assist in keeping the temperature. Be careful, as heat always drops when chicks are taken out. You can have a small glass door in front of egg-drawer, to observe thermometer, if desired. Always change position of trays when eggs are turned, putting the front one at the rear.

CHAPTER VIII.

CHICKENS ON THE FARM.

Generally in calculating the profits of the year, the farmer overlooks the chickens entirely, and very often it is the case that if he should think of them at that moment there would be very little to add to his account as their contribution. The reason for this is that they are neglected throughout the whole year and cannot be expected to do much; nevertheless, if an account had been kept, even with this neglect it would be found that many a chicken had been served up as a palatable meal during the year and many dozens of eggs gone to the store, besides the large number used at home. And this without any attention except to throw out a little ice-cold corn winter afternoons and perhaps some slop (such as fed to the hogs), while there were little chickens; for this young and old scrambled, with the result that little ones were often trampled upon and killed. The hens stole their nests and dragged their chicks around in the wet and scratched what food they could for their young; if half the brood survived they did well. Later in the summer they roosted in the trees; when the snow came it drove them to seek shelter in the stable, granary, machinery or carriage shed, where they left their marks.

If, perchance, shelter was provided for them, it was an old shed with a leaky roof and no light, except what came in through the door and numerous cracks, which also served equally well for admitting the cold. Here the droppings were allowed to lie from one year's end to the other, and the dead chickens, of which there were apt to be several in the course of a winter, stayed just where death overtook them until warm weather made them so offensive that they then were dumped back of the pig-pen If a hen happened to want to sit in such a place, eggs were put under her and then she was left to shift for herself; the other hens helped break her eggs and the remaining ones were left in a soiled condition; occasionally she was yanked off by the tail, and if there were more eggs under her than originally, a few of the freshest looking ones were taken away.

If the hen chose (as was most likely) some more agreeable place for depositing her treasures, they were gathered whenever chanced upon, unless it was very evident she had been sitting for some time. If, when hatched, the mother and brood were cooped up, they were fed at very irregular and uncertain intervals.

The older fowls were allowed to roam at their pleasure, and often annoyed the cattle, and scared the nervous horse by picking away under them or in their feed troughs, until the more timid animals stood trembling while their feed was devoured by the hungry feathered bipeds. A favorite place for sunning themselves was the back doorstep of the house, and this was a source of much annoyance to the good housewife; her cherished garden was scratched up by them and generally she concurred with her husband in considering the chickens a nuisance, which for some not clearly defined reason had to be tolerated on the farm, instead of a source of much profit as they might easily become.

Dear reader, this may not be the condition of the chickens on your farm, I hope it is not; but is it not a fair, if not actual representation of the manner in which some farmers of your acquaintance treat their biddies?

IMPROVING THE STOCK.

I would not, for a moment, have you think that I believe it is absolutely necessary that a large sum be expended on costly houses, incubators and fixtures, in order to raise chickens in a civilized and profitable manner. You will see by referring to the chapter on houses that comfortable quarters can be provided at a very moderate cost, and unless very early chickens are desired the incubator can be dispensed with Thoroughbred stock is not necessary in order to get good laying hens or good chickens for the table or market. Common hens well cared for will pay better than pure-bred ones that are neglected. Care counts more than blood alone any day. It is well always to select the best hens of your common stock for breeders, and if possible get a pure-bred rooster of one of the breeds that possess the qualities you wish; if you are keeping chickens for their eggs, select those of your hens that you know are the best layers, and mate with them a rooster of one of the breeds noted for their egg-laying quality; if table fowls are desired mate with that end in view. By continuing this process, changing roosters from time to time to avoid the evils of in-breeding too closely, in a few years you will possess a flock that will nearly equal pure-bred fowls for practical purposes.

If you do not feel justified in purchasing a thoroughbred cock even, you can by this same method of selection in breeding stock greatly improve your flock when using a male of the common run. But be sure to give your chickens the same care as you would were they high-priced stock.

In buying pure-bred poultry it is not necessary to purchase the highest scoring ones. Often there are birds whose blood is just as pure as the purest, and as good for all purposes, except the show-room, that can be obtained for one-half, and often much less, the price of their most noted brothers; these are the birds we advise farmers to buy, unless they wish to raise show-birds.

Probably the more popular way of getting pure-bred chickens is to buy sittings of eggs, and in this way the best blood of the country is obtained at a much lower figure than by buying birds, but, of course, it delays the improvement of your flock one year. In having "pure-breds" to start with, you not only obtain the desired results soon, but also have eggs and birds to sell, for which your neighbors ought to be glad to pay more than the market price for culinary stock.

CHILDREN AS CHICKEN-RAISERS.

Many a man who takes first-rate care of his stock seems to think that chickens are "too small fry" for him to give any attention to, so they are neglected or turned over to his wife and children. Now, in many ways a woman can give better care to chickens, little chicks in particular, than a man, but it is hardly right to ask her to attend to the hens in midwinter or at any other time, when there is a large family of children to care for and several hired men to cook for, especially if there is no help in the kitchen, as is too often the case. Children usually take great interest in chickens and like to care for them, especially if a few are given them to be "their very own." I would recommend that fathers give each child a hen or more and let them have all the chickens she raises, also the money they bring for their own; in this way the children are not only led to show greater zeal in their work but also learn to save their money as they do not when some is given them which cost no labor on their part. Nevertheless the chickens ought not to be turned entirely over to the children, because they may forget some things that will make a material difference in the profits accruing from the hens.

POULTRY-HOUSES ON THE FARM.

In building a house for your poultry place it on a high, dry spot and have it somewhat sheltered from the cold winter winds if possible; have the front of the building to the south and be sure there are plenty of windows in it. It is best not to have it connected with the stables, as the fumes from them are more or less injurious to poultry; and if they, through neglect, become lousy, the cattle and horses will be almost sure to be affected.

As poultry can be kept more cheaply when running at large than when confined, it would be better to put the chicken house at some distance from your own dwelling, so that they will not litter up the stoop or doorstep. If they have good quarters they will not be likely to bother much if table scraps and food are not thrown out of the back door to attract them. To save all risk a light fence of lath or wire netting may be built about the dwelling.

When you have your buildings and breeding stock you are ready to begin business. Do not expect eggs too soon if the fowls have just been bought, as moving usually disturbs them, and it takes some time to become accustomed to their new quarters. If your building does not contain arrangements for separating the sitting hens from the others, take pains when a hen is set to fasten her in with laths so that the others can not disturb her, and when you let the sitting hen off to feed be sure that all others are shut out of the house or some may enter the nest to lay, and the "sitter" on returning make such a row that the eggs be broken or she go on another nest, and when the laying hen comes off the eggs become cold. Where several hens are set in a row of nests all alike, some advocate painting them different colors to assist the hens in finding their own nests and they assure us that they have much less trouble with hens getting on their neighbors' nests than when there is nothing to tell the nests apart.

Sitting hens should be allowed to come off every day, but do not disturb them if they do not seem inclined to leave their nests. Some hens will not feed oftener than once in three days. Have plenty of grain and fresh water at hand and a box full of road dust or sifted coal ashes for them to roll in; this helps to keep off the lice. If you leave while the hens are off, do not stay away very long, as some may have to be put back on their nests. While sitting, and just before hatching time, dust the hen and nest well with insect powder or Scotch snuff as a safeguard against lice. It is better to put a sod or some dry earth into the bottom of the nest, and cover this with cut straw or hay, or many prefer very fine shavings, as with them there is no danger of the hen scratching for the seeds. Have a sponge and some warm water to wash off any eggs that may be soiled by others being broken, for if left daubed up the embryo chick may be smothered; at any rate the odor of the broken egg will not be agreeable if allowed to remain in the nest or on the eggs. It also becomes a breeding bed for lice.

When hatched, allow the chicks to remain in the nest until they are all dried off, but if the weather is cold or the eggs hatch slowly, put the first ones behind the kitchen stove and look out for gentle pussy.

As soon as the chicks are all dried off remove them with their mother to a roomy coop placed in some dry spot. After twenty-four hours feed as suggested in Chapter II of this book. When they are a few weeks old they should be allowed to forage for insects with their mother on pleasant days, but she should be shut up every night, otherwise she will lead the chicks off through the dew in early morning, and they may become chilled and die; on rainy days also keep them under shelter. If any are caught in a shower and get soaked treat as prescribed elsewhere in this book.

Let the old hen stay with her brood as long as she will; when she leaves them they will probably spend their nights in their old coop if left undisturbed. If the chicks are quite young when forsaken, and the nights are chilly, throw an old carpet over their coop. Let them "roost" on the ground as long as they will; roosting on perches while young is apt to make their breast bones crooked. When the nights become quite cold or when the chicks begin to perch upon fences or trees, gently remove them, after dark, into the hen house. Any that get away while you are transferring them are not to be chased about; let alone until the next night. Very likely part of those that were carried into the hen house will be in the tree again, if so, take them in once more; in a few days they will become accustomed to their new quarters and go in of their own accord.

WINTER CARE OF CHICKENS.

After the ground is frozen, chickens ordinarily cannot find as much to eat as during the summer, and their feed should be increased accordingly. During the winter a part of the house should be penned off by a board from six inches to a foot in height, and this space filled with straw, hay or leaves; into this their grain should be scattered; in scratching for it the fowls will get much needed exercise while confined. If something of this kind is not provided they will become too fat and eggs will be a scarce article. Where there is not sufficient room in the house for this "scratching-place," roof over a place on a protected side and put the straw under this. Be sure that the sun can shine on the straw and do not put the roof where it will shut off the light from the house. A dust bath also ought to be provided. It is a good plan to gather several barrels of road dust in the fall and keep it in a dry place; put a box of it where the hens can reach it; replenish this from time to time. Sifted coal ashes are very good for the same purpose; the fowls relish the bits of coal and clinkers which they find in the siftings.

During the winter give the fowls warm water at least once a day; if your building is so cold that in extremely cold weather the water is apt to freeze if left standing in the vessels, it had better be thrown out after each meal; in this case they will need refilling three times a day, for chickens need water in winter as well as in summer, and "eating snow" will not take the place of fresh drinking water.

On warm, sunny days the hens greatly enjoy being out of doors, and if there is deep snow on the ground a space in front of their house may be shoveled clear for them. For the best feed during this season of the year see the chapter on "Care and Management."

BROILERS ON THE FARM.

If you are located on a line of railway offering good transportation facilities to a large market it might be advisable to purchase an incubator and brooder and raise broilers. In this case some of the earlier pullets must be kept for the next year's layers or the eggs will not be forthcoming from which to hatch the early broilers. Most of the breeds are mature enough to begin laying when six months (often earlier) old, if they are well cared for, but they will not produce as strong chicks as those which are more mature. Old hens can not be depended upon to lay early.

As a rule, where eggs are the object, it is best not to keep hens after two years old, and many advocate turning them off the summer after they are a year old; during July and August they usually command a good price.

If it is not desirable to raise broilers, the chicks need not be hatched as early; but for winter laying the bulk of them should come out in April and May. Hatched at this season they can be raised at less cost than the winter chicks. Turn the cockerels off during the summer and early fall, or caponize them. Do not have a flock of poultry composed of from one-third to one-half old roosters, that are worse than useless about the farm, and bring a very low figure on the market. The surplus pullets may be turned off with the cockerels as spring chickens, though they will sell at a good price at almost any season of the year.

An incubator will be found a profitable investment where poultry is raised on a large scale, as in that event the hens are not needed for hatchers and will usually get back to laying much sooner than when allowed to sit.

Although late hatched chickens can not be as profitably kept for layers they will add largely to the profits as market birds, and many poultry raisers keep on hatching throughout the spring and early summer.

Ten to a dozen hens to a cock is a fair proportion, and even more of the lighter breeds can be safely kept. If you have a trio of thoroughbred fowls and wish to keep their eggs separate from the rest of the flock, it is best to confine with them a few hens whose eggs are a different color, otherwise the male may be too attentive, and the hens stop laying.

Persons living in towns or where, if they raise chickens, they must keep the fowls confined to limited runs, will find them a source of pleasure as well as profit, if they are given careful attention, especially in providing plenty of green food. Nothing can be more satisfactory than fresh eggs or a fat fowl of "our own raising" to the family in the village, and the neighbors are always willing to take the surplus eggs and chickens, for they know what they are buying in such a case. Even in a large town, where the house lots are very small, chickens may be kept, but there it is best to have the heavier breeds, such as the Asiatics, for they are more docile by nature and bear confinement better than others; or Bantams may be tried; as a rule, they are very prolific layers and take up very little room. Besides furnishing that rarity of the city, fresh eggs, they will yield as much enjoyment to a common mortal as a circus, and the children will amuse themselves for hours at a time in watching these cute little pets. A half dozen of these pigmies could be kept on the back porch of a third-story flat.

Those who say poultry does not pay do so because in the first place they do not expend the proportionate time and brain in caring for their fowls that they do with their other stock. In the second place, they do not keep an account, hence the many little sums are overlooked when compared with those derived from the cows, for instance, where many times the capital is invested. Take care of your hens for one season, credit them with all the eggs and chickens used at home as well as those sold, ot course charging the feed and time to them, and see if they do not yield a greater profit proportionately than the average products at your disposal.

Those who have thoroughly tried it, either as a business in itself or as a side issue, are almost unanimous in declaring that it is one of the most profitable, if not the most profitable, branches of farming. One must not expect to do well at it unless he is willing to devote time and talents to it and even then there are a few who, despite their best efforts, will fail; the same is true in every business and profession. To such I would say try something else, but to the average man I would recommend keeping at least a few fowls, if situated so that it is at all practicable to do so.

EGG FARMING.

In these days of specialties, farming is much more divided than formerly; we have fruit farming, dairy farming, duck farming and egg farming. Of those who have attempted the production of eggs on a large scale a few have succeeded. The chief cause of failure with those who have not been successful has been that they seemingly undertook it with the idea that all they had to do was to multiply the small flock with which they did well by 100 or 1,000 to get 100 or 1,000 times as much profit. Multiplying the number of hens is the very smallest factor of success in egg farming.

The first element in successful egg farming is small flocks; the second is economical housing; the third, facilities for handling and cheaply caring for the flocks. The flocks being small each one can be cared for and treated as though it was the only one on the place. The houses should be warm and convenient for the birds and the attendant. It will be much cheaper to limit the range of the flocks by their own free will instead of by fences or runs. To do this, place the houses at short distances apart—ten or twelve rods—and have them of different colors and with different surroundings so each may be readily distinguished by its occupants when they wish to seek "quarters" at night or for laying. One of the most successful egg farmers in this country arranges the houses of his hens along streets, ten rods apart, through his farm. Half of the 120 acre farm is used as a hen yard one year and the other half cultivated. The next year the houses are moved over to the cultivated half, and the hen yard used for cropping purposes. The houses are built on sills which form runners and a good team is used for moving them in the spring. During the summer they are moved their length every week; in this way the houses are kept sweet, as the soil where they are to stand is plowed before the houses are moved. The object of this is to keep the houses healthy with fresh soil as an absorbent of the droppings which renders the air pure without the labor and expense of a daily cleaning, and also "fixes" the fertilizing elements in the droppings and enriches the soil for next year's crop.

When the flocks are left to make the limit of their range, one thing has to be guarded against and that is their following the attendant who feeds them from one house to the next and so on till there is an intermingling that it would be hard to separate. To prevent this running to the neighbors', the fowls never see the attendant when food or water is given them, this being done at night or while they are shut up so they cannot see how the supply is furnished. If the range is limited by fences this precaution in feeding need not be taken.

In winter the air of the houses is kept pure by daily covering the droppings with dry dirt that is stored in the summer or fall. In the spring the building is pried up from the mound that has been formed, and, after being drawn away, the mound is plowed up and scattered with a scraper before the "yard" is plowed for planting.

The most successful egg farm will need at least three sets of hens unless incubators are used for hatching purposes. One set will be the sitters or mothers, one the breeders and the other the layers. The breeders should be of the best laying strains; these with their male companions should be kept somewhat apart from the layers and fed for hardy, vigorous offspring, not crowded for the greatest possible number of eggs within the first few months of their laying days. The sitters will be of some of the heavy breeds or crosses therefrom; they should be fed so as to be through laying in time to be ready for incubation in late winter or early spring and then right along till June. They should be about half and half, hens that were used the year previous in rearing two or more broods each, and the earliest hatched pullets of the previous spring.

If the poultry farm can have a soil naturally well drained and slightly sloping towards the south or southeast it will be an ideal location if it is near a large city where customers can be had and supplied regularly.

On a farm managed as suggested above, half used as a hen range one year and cultivated the next, there will be need of providing temporary shade and shelter. This may be afforded by rough board sheds, two or three feet high in front and slanting back to the ground. Or a frame may be made of poles and covered with brush, straw or corn stalks.

When engaging in egg farming it must be borne in mind that it will be almost absolutely necessary to supply animal food in much larger proportion than where a few fowls are kept, as insects abound where vegetation thrives, and there will be but little vegetation in a hen yard that has a house every ten or twelve rods and on land that was plowed and cultivated the year before. For the same reason a good supply of green stuff must regularly be provided. To this end the crops raised on the balance of the farm may be largely such as will supply green and succulent food, such as cabbage and turnips. A partial supply of the green food necessary may be given by sowing lettuce, oats and other quick-growing greens in strips among the houses and protecting with wire netting or other light fences, while getting a start, then let the fowls have access to these patches while others are being protected. All this means work and constant work. No one need expect to succeed at egg farming without work, any more than at any other kind of farming. Do not undertake it unless you are willing to work and better not undertake it at all except by growing up to it from a small beginning.

CHAPTER IX.

HOUSES AND FIXTURES.

"I am thinking of building a new hen-house for a flock of from a hundred and fifty to two hundred. What do you consider the best size and plan for such a building?" "Please give plans for substantial but economical hen-house." "How would a sod house plastered and whitewashed do for poultry?" are samples of questions that come to my desk. In answering them the section of country from which they come must be considered as a factor in the construction of the buildings. There are two or three things that must be considered in building every hen-house. They must be dry, warm and large enough for the flock to be kept. All other matters in construction and ornamentation may be left to the taste and means of the builder.

The necessary cost of the poultry house will depend very much upon the price of lumber and other materials in the locality where built. To this must be added the wages paid for labor. The latter expense will be saved by a great many who keep poultry, especially those who keep a small flock upon a farm or in a suburban village, as they will do the work themselves. As examples of the range in cost of buildings we mention one in Massachusetts for one hundred hens, where lumber and wages are high, that cost but a few cents less than $100. Another in the same State, to accommodate 500 hens, cost $400 where the owner did most of the work himself. In Ohio, houses that would accommodate fifty fowls each were built at $30 apiece where the labor was paid for as well as all material bought. "Out West" a house for 100 was built with an outlay of only $10 for material. This was the only money outlay. The "material" bought was a few rough boards, nails, glass and window sash. The work was done by the owner, and the other "material" that entered into the construction of the building was poles, sods and straw. This house may not have been called "handsome" from an architectural stand point, but it was very "handsome" if "handsome is that handsome does," for it was warm and the occupants laid right along through the severest winter weather. Give hens "warmth" when the mercury is playing around below zero twenty to thirty degrees, and they will talk merrily and rejoice your pocketbook with high-priced eggs regardless of the lack of beauty and "lovely" surroundings that adorn their quarters—proper feed, of course, being given them. All the beauty of outside finish and architectural symmetry will not produce eggs without warmth. Bear that in mind always.

We give a few illustrations, plans and specifications of buildings that have been built. From them, with such modifications as locality and circumstances indicate, suggestions may be derived as to what may be done in the way of poultry houses.

As to fixtures and furniture, we believe the fewer the better and we would have all loose so they may be readily and quickly removed from the house to facilitate cleaning and keeping free from vermin. A few flat perches, movable boxes for nests, a dust box, one for gravel and grit, a V shaped feeding trough and a drinking vessel are all we consider necessary. If milk is used have two drinking vessels. Have those that can be readily cleaned and scalded. This applies to feeding troughs as well as drinking dishes. If there is a covered shed connected with the house or if the perches are above a second floor the dust and grit need not necessarily be in boxes. If the perches are at all elevated there must be easy approaches to them.

The idea that the whole of the south side of the poultry house should be of glass is now discarded by many of the best poultry raisers. All hold that there must be plenty of light and quite a liberal amount of glass. If there is much glass the strong, dazzling effect of so full a flood of sunlight may be modified by whitewashing or painting the glass or part of it. In the summer time the sash may be removed and wire netting or lath placed over the window spaces.

For winter warmth there should be board shutters to put up at night.

A FIFTEEN-DOLLAR HOUSE.

We are indebted to Mr. J. R. Brabazon, of Delavan, Wis., for the cut below and the following description of his cheap poultry house.

The poultry house is 12x16 feet, five feet posts and nine feet at the highest rise in the roof. Faces the south and is

intended for three breeding pens, or can be turned into one house. Roosts are all movable and nests too. The windows are four feet long and two feet wide, and a window to each pen. Is made two boards thick with paper between, and the pens lean as is shown in the cut as a shingled roof. The windows have an angle of two and one-half feet from top to bottom to give plenty of heat in winter, and I keep my Leghorns in it, and in this latitude of some 30° to 40° below zero I have never had one freeze its comb yet. It can be made here for $15 if a man has ingenuity enough to build it himself, but where lumber is higher than here in Wisconsin, it may cost a little more. Any old boards can be used for the first thickness. Don't use any 2x4 only in the corners and roof two feet apart, and run the first boards lengthwise, the next up and down and batten, and you have a frost-proof hen-house if you put the boards close together in roof and paper before shingling. It will accommodate forty to fifty laying hens and more when in the fall young stock is in. Never crowd your fowl house.

AN INDIANA POULTRY HOUSE.

Mrs. A. L. Smith, Princeton, Indiana, thus describes a poultry house that is very convenient where more than one breed is kept, or where the poultry has to be confined:

The poultry house is 12x40, divided into five runs, making them 8x12 in the clear. It is on the hillside facing the south, with a good-sized window in each run, with one additional in each end of the house, east and west.

A brick foundation with 2x4 studding laid on top, then studding set up, six feet on north side and ten on the south, wide enough apart to tack building paper on the outside, and tarred paper on the inside, then siding joint on outside and ceiling inside, thus making a double hollow wall with two ply of paper, which entirely protects the fowls from frozen combs. I lined the doors with tarred paper and then a layer of ceiling. This, with close window shutters on outside, makes a very close room and proof against freezing.

For ventilation two tubes made with foot wide lumber, within one foot of floor and up through the roof, and to which is attached a nice galvanized iron flue, to finish it nicely on top.

For admitting fresh air, but not cold air, I got three hard burnt straight tiles, and three elbows, also three tin spouts or tubes six feet long, and some common three-inch tile. Beginning a few rods out from the house, with the hard burnt straight tile at the surface, I run the common tile under ground till it came within the house four feet from the south wall; at each one of the partitions attach the elbow and bring it to the surface, then insert the tin spouting, and you have an inlet for pure air, coming a few rods under ground, and sufficiently warmed, so it will not cool off the house. The tin spouting carries it above where the fowls roost, so they get no draft. I find when my shutters are all closed, by putting the hand over the top of the spouts, I can feel the fresh air rushing in, but it is not cold, because the ground has warmed it as it passes through the tile. There are three of these ventilators.

The roosts are made by taking a twenty-four inch board for a table, putting slanting edges on six inches deep, and putting on each end two uprights, one foot long, with two inch slots cut in upper end, into which the ends of the two-roosting poles rest. These perches are two by three, rounded on top. This wide trough can be swept off often, and the droppings carried out, as it sits on two little trestles about eighteen inches high. I raise the ends of perches occasionally, and pour on coal oil, and so I never have mites in my house. Keep perches and dropping-trough away from the wall.

The house is covered with tarred paper and after six years of use, we re-covered with iron this paper helps to keep the house warm in winter. Window-shutters closed in winter at night and windows all open in summer with wire screens to keep out vermin.

The partitions are boarded up about three feet, and wire netting from that to the ceiling or roof.

Small doors at bottom leading out of each run, into a large grass run, from one to two acres in each, except middle one which is left, and used for a cockerel run or later in the season for the first brood of chicks, when they are old enough to remove from the brood house. In this run I have a cool roosting coop made of inch square oak laths nailed on close enough to keep out rats, with a wire screen door; the air can pass through in all directions and it is under the shade of a tree, size 4x10 feet. Then when cold weather comes we take them from these outdoor coops into the house above described.

The cost of the house was about $100, but it accommodates five varieties of fowls; one run need not cost over $25. Sometimes persons may find windows and doors that carpenters have taken from old buildings, that are just as good for poultry houses as new ones, and so make it still cheaper. We put on a coat of paint and in two or three years give it another and so will always have a good house. The doors leading from one run to the other all have strong springs so they can't be left open, and the gates from one yard to another with heavy weights to close them. In this way there is no danger of having them open.

The ceiling on the inside may be of any cheap lumber and put on rough, as it holds the whitewash better. We apply a thick coat of whitewash in which carbolic acid has been mixed. This is good for the health of the fowls, and death to lice, should they happen to get started. But this with the tarred paper never has let them start with me, and it has been eight years since it was built, and has been occupied by about 100 chicks all the time.

A MISSOURI POULTRY HOUSE.

Mr. Elmer Putnam, Sheldon, Mo., thus describes a cheap poultry house in Poultry Topics:

I use nothing heavier than 2x4. First, I nail the sills and joists together, then lay the floor for a room 8x10 feet, then cut two posts two feet ten inches long for the low side; toe nail at the corners, put a plate on top; this makes the low side of the house three feet high. For the front I take two posts four feet ten inches long; this makes the front when plate is on five feet high. I have from two to four windows in each house, one sash each. I have from two to four openings at the bottom underneath the windows. I have no ventilator for I have never needed one. I have one pair of rafters at each end with a ridge pole at the peak; the rafters for the long side are sixty-seven inches long; on the short side, the rafters are thirty-three inches in length. In building I use a board long enough to make one length for each side of the roof and for the low side of the building.

The windows are removed in summer and wire cloth or netting used instead. Such a house 8x10 can be made at from $10 to $14. Two of my houses have open sheds at one end. I use anything for roosts I can get. Four-inch boards cut to six-feet lengths do very well. My roosts lay upon low, short benches about two feet from the floor. I like short benches better, for the reason I can with the short roost boards put a row of roosts on each side of the house and across one end; this leaves an open space in the middle to work in when I powder my hens at night.

The floor should be tight so that nothing can go through· Everything inside should be loose so that it can be removed easily.

Part of the nests are made of wire netting and part from boot boxes. The netting I used was eighteen inches wide and it takes a little over four feet in length for a good nest. One side I slit up some six or seven inches in four places, then make two hoops of number 9 smooth wire to fit snugly inside after ends of netting are fastened together, one hoop at bottom, one at top; the portion that I slit up I bend in for a bottom, then I make four braces the height of my basket, fasten one end on bottom hoop, one on the top, being sure to have them long enough to draw the netting very tight, these braces should be equally divided around it. Then put two short cross pieces of wire under the bottom with two hooks to hang it up and you have a nest that has cost about five cents, and when cleaning one instant in a flame and the wire is almost red hot, thus destroying everything on them.

The boot boxes I remove top and one side and enlarge them so that the nest is about sixteen inches in all directions. In front, a four-inch strip holds the nest in while the back part of the cover I leave loose so that I can get the eggs.

These boxes are on feet about one foot high; when in use the open side is next the wall so the hen must go in from behind. Then when I set a hen in them the box can be set up close to the wall and the hens are shut up from the laying hens. Then in summer they can be carried out in the shade in any out-of-the-way corner; they are up off the ground where nothing can get in them without climbing and are light and easy to carry.

HOUSE FOR ONE HUNDRED HENS.

Fig. 1 (page 210) shows a good house for 100 fowls. It is 50x12 feet, which gives five rooms 10x12 feet, four of which can be used for fowls, while the fifth can be used for storing food, etc., sitting hens in season, and if desired, a stove for cooking food may be set up in this room. In case a stove is desired, have a chimney in the center. There is no window in this center room, but it is sufficiently light, for the partitions are of wire netting down to within two feet of the floor. If more light is desired, a window may be put in the upper half of the door. The house is set on a stone foundation, and in this foundation openings are left for the fowls.

FIG. 1—HOUSE FOR ONE HUNDRED HENS.

One of the openings is in the west end and does not show in the illustration. Ventilation is secured by the two box ventilators. Roosts and nests arranged to suit the fancy of the owner.

Fig. 2 (page 211) is a neat little poultry house twenty feet long, eight feet wide on the ground, six feet high in the rear, and six and a half feet in the roof. It is built of matched and dressed lumber, battened and painted. The frame is 3x4 joists, lathed and filled in with sawdust on all sides and roof, then plastered; gravel bottom; three windows of twelve lights, 9x13, and a small window in upper half of door to admit the morning light. The nests are on the ground, under the windows. This house proved a success in a severe winter, the thermometer indicating only 3° below freezing when it was 26° below zero outside.

Where the climate is not too severe a similar plan might be followed, but instead of the lath and plaster, double boards

with building paper between might be used both for sides
and roof.   A house like this may be extended to any length
desired, using wire netting or lath for partitions to separate
the different flocks.   In localities where the cold is such as
to freeze water solid in the poultry house it is about impossible to keep the combs of single-combed Leghorns, Minorcas
and such breeds from freezing unless provisions are made
for artificial heat.   We do not think it advisable to
keep the poultry house up to hot-house heat, but we know
from experience that one cold night will stop all laying for

FIG. 2—A WARM HEN-HOUSE.

a month or six weeks where we depended on Leghorns to
produce the eggs.   A stove and a few cents' worth of fuel
one night would have saved our hens pain and disfigurement and given us dozens of high-priced eggs.   A kerosene
heater would pay for itself in one night of such weather in a
house where twenty or thirty single-combed Leghorns are
roosting.

A CONVENIENT POULTRY HOUSE.

The full-page illustration (Fig. 3, page 212) shows the elevation of a poultry house that is built with the back part in
front, so to speak, and has the windows of two sashes in
the lower side, the front, which should face the south.   This
arrangement has proved very satisfactory to the poultry
raisers who have tried it.   A house like this can be built any
desired length, and divided into different sized rooms to suit
the convenience of the owner.   It should be six feet high in
front, and ten or twelve at the rear, according to the width.
A passage-way three feet wide extends the whole length of

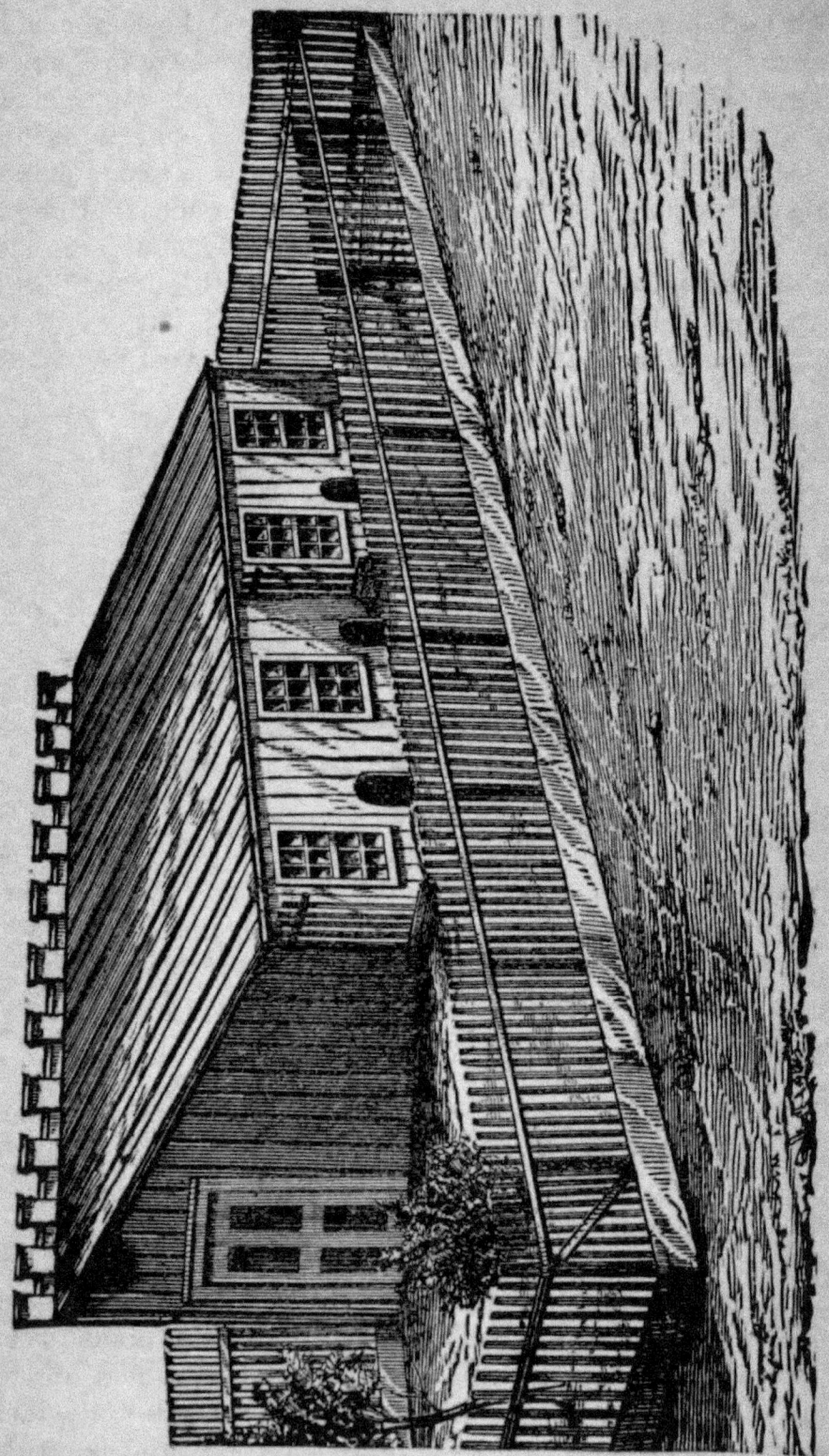

FIG. 3—A CONVENIENT POULTRY HOUSE.

the building. The windows are hung with weights, and by a simple arrangement of cords and pulleys can be raised or lowered by pulling a cord in the alley. The space over the passageway in one house like this was used for storing chicken coops when not in use. In another it was finished with a tight floor and used as a roosting place, the fowls "climbing to rest" by means of a ladder which will be described farther on. In still another there was a pile of refuse hay, and a few nest boxes scattered in the "loft," and how the hens did like to lay up there.

A house after the same general plan as the above has been built at an expense of twenty dollars for material. Following are the dimensions: Width, twelve feet; length, twenty feet; height of front, seven feet; height of back, five feet. The material used was 650 feet foot boards, 130 feet 2x4 scantling; four windows, two sashes each; 260 feet of roofing paper, one door made of boards and battened, 300 lath, seventy-five feet of two-inch battens and twenty pounds of nails. There was no battlement or trimmings of any kind, and the twenty dollars, as stated above, paid only for the material. Such a building would not do for winter quarters in a cold climate unless the sides were well protected with outside walls of straw or manure banking. We have built two after the same plan, 10x16 feet, nine feet high in front (this was the front facing south) and six feet high in the rear, with a roof of No. 1 shingles, at a cost of twenty-five dollars each, reckoning the work at $1.50 per day. This included one partition and nest boxes in each. We secured ventilation by extending the roof rafters (2x4s) six inches beyond the top plate in front and facing them with a ten-inch board. In this way fresh air entered without admitting the storms. In very severe weather straw stuffed in the opening between front of building and the face-board closed the ventilator and kept out the cold.

AN OHIO POULTRY HOUSE.

Mr. A. C. Pepron, in the Ohio Farmer, describes a house which he planned for 100 Brown Leghorns:

Building stands east and west, so as to face the south; sixty feet long, twenty-five feet wide; foundation of brick or cobble stone, two feet above ground and one foot in the ground: roof one-third pitch, making the building ten feet three inches high in the center; door in each end, and over the doors a small door for ventilation. In the south half of the roof put at least ten windows, one-half at peak, with the others at or near the eaves. The upper might be put in flat with the roof, and the lower would be better as dormer windows, and rise from the foundation, but this would make the cost considerable more than to put them in with the pitch of the roof. The ends and roof should be made double, with at least a two-inch air space, and also a sheathing of tarred paper between. Of course this makes the cost much more, but it will pay in the end; the fowls will not feel the sudden changes of temperature in winter, and unless there is long continued cold weather, the temperature of the air in the building will not go down to, or much below, freezing. The outside of building should be banked up all around as high as the foundation wall. To do this I would draw dirt and grade it up and make it permanent. Put a ventilator in the center, extending at least four feet above the roof and two feet below into the building; it should be sixteen inches square, and over the top and about four inches above a roof or cover, so as to keep out the snow and rain. The ventilator should be kept open all the time, and the end ventilators over the doors open only when necessary. This ventilator in the center can be used for a chimney by running a stove pipe up through the center and having a galvanized sheet iron top above the ventilator. The doors should be double and fit tight.

HOUSE FOR LEGHORNS.

Fig. 4 shows another man's (S. L. Roberts, of Nebraska) notions of a house for Brown Leghorns. Mr. Roberts says: "I have a superb pen of Brown Leghorns, with extra fine, large combs, and although our thermometers registered 15° below zero, my fowls all came through the winter untouched by frost. I have two buildings exactly alike, each eight feet wide, sixteen feet long, eight feet to top of square in front and four feet behind. These pens stand ten feet apart, with a roosting coop between. The roosting coop is four feet wide by ten feet long, two feet high in front, and one foot high behind, with a two-foot pit or cellar under it with roosts on a level with the ground. Two trunk ventilators run down within six inches of the ground, with a division in the middle. This coop answers for both pens. The cost is trifling; three twelve-inch boards ten feet long and the roof complete the building. A hole cut into this from each pen and it is finished.

"I can heartily recommend this roosting coop to all breeders of Leghorns for winter. The advantages are many.

FIG. 4—HOUSE FOR LEGHORNS.

It is warm in winter and cool in summer; and it is difficult for night prowlers to secure their booty. By this plan your poultry house can be kept clean; you will have healthy birds and plenty of eggs. You can close it up when you give the morning meal. Fowls will have no chance to learn the habit of loafing on the roosts half of the day. The objection to glass fronts is overcome, as the cold does not reach them, and no one need pity me when he sees one hundred and fifty panes of ten by twelve glass in a house eight feet by fifteen feet. No fancier should have his fowls roosting in the daily quarters when it is so easily remedied with so many advantages.

"The best and cheapest material to use for building is twelve-inch hemlock flooring tongued and grooved, which costs from $16 to $18 per thousand, according to location.

This used as siding should be put up and down. It also makes a good warm roof covered with two-ply tarred felt. I have the roof and north end of pens covered with three-ply tarred felt; the balance, with the roosting coop, is covered with two-ply felt. Inside of the pens everything is neat and clean, and the air is fresh and sweet, thanks to the roosting coop and trunk ventilators."

HOUSE FOR SEVERAL BREEDS.

Fig. 5 shows a neat-looking poultry house arranged for the keeping of several different breeds. It will answer just as

FIG. 5—HOUSE FOR SEVERAL BREEDS.

well where only one is kept in several flocks. The middle room is used for a store-room, and is furnished with a stove for cooking feed and warming the building in severe weather.

A HOUSE FOR SEVENTY-FIVE FOWLS.

Fig. 6 shows the elevation, and Fig. 7 the ground plan of a poultry house fifteen by eighteen feet, that will meet the requirements of the poultry raiser who desires to winter from fifty to seventy-five fowls. A represents the laying-room; B, the roosting-room; C, the room for feed and sitting hens, and D a bin for grain. The nest boxes shown in the partition between the laying and sitting rooms are intended to slide back and forth. A house like this may be built any desired length, and a continuous passageway made by changing the arrangement of the roosts. Instead of having as much glass in the front as shown in the cut, some prefer

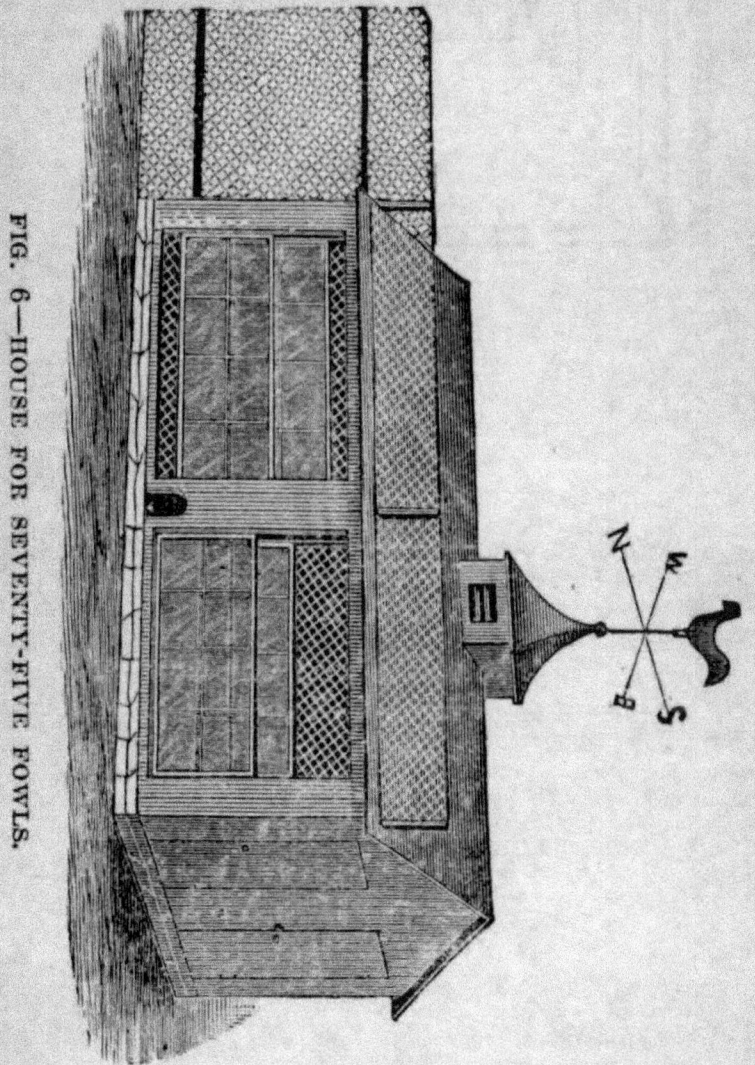

FIG. 6—HOUSE FOR SEVENTY-FIVE FOWLS.

two fair-sized windows and a small four-light window in the upper half of each door. The same inside arrangements may be had in the shed-roof building, Fig. 3.

Fig. 8 shows the elevation of a house for one who desires to winter a good-sized flock of hens, and have plenty of room for raising early chickens.

Fig. 9 shows the elevation of a very convenient poultry

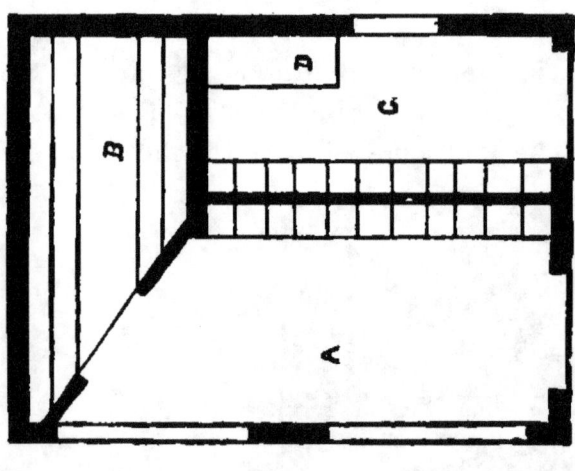

FIG. 7.

FIG. 8—HOUSE FOR FOWLS AND EARLY CHICKS.

FIG. 9—A CONVENIENT HEN-HOUSE.

house which can be built ten or twelve feet wide, and any desired length.

Figs. 10 and 11 show the ground and end view of the inside arrangements of this house.

A BARN-CELLAR POULTRY HOUSE.

One of the most comfortable and generally satisfactory poultry houses ever brought to our notice was one-half of a

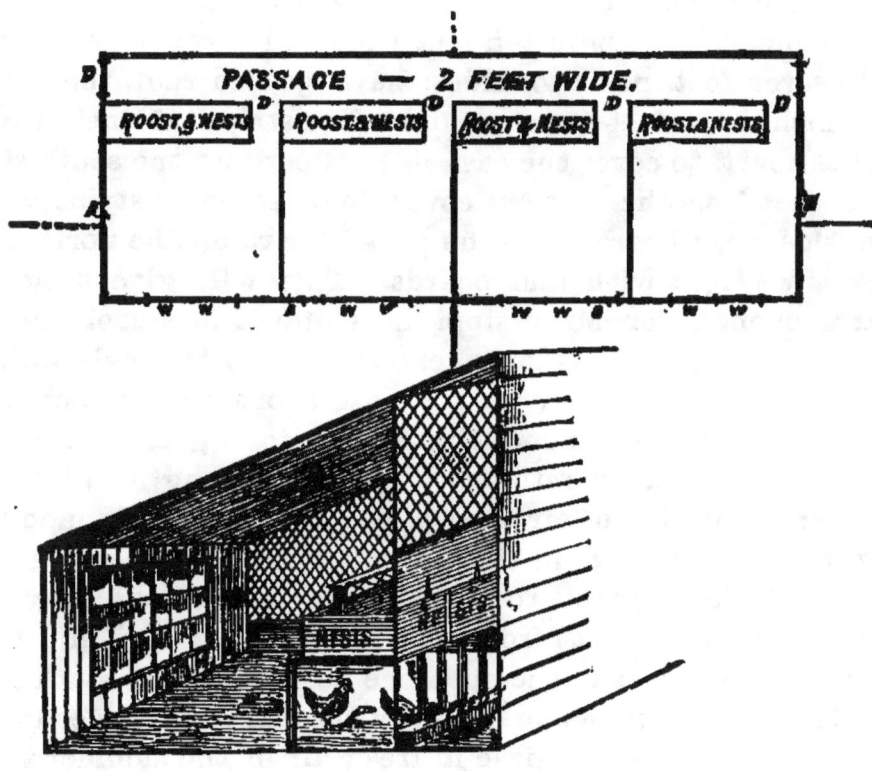

FIGS. 10 AND 11.

barn-cellar. The barn was built on the south side of a hill, and the cellar walls were of stone. The inside of the half that was used for a poultry house was finished off with matched boards. There were two large windows in the front, with board shutters to close over them winter nights. A box ventilator ran up through the barn floor into the room above. A door opened into the manure cellar, and in cold weather the fowls spent the greater part of the time during daylight scratching in the manure piles, and scratching and loafing in the barnyard sheds. When the mercury was down to twenty-two degrees below zero outside, it was above the freezing point in that cellar.

A MISSISSIPPI POULTRY HOUSE.

In building poultry houses in the South, summer comfort as well as winter warmth must be sought; in fact, unless care is taken the winter warmth will be more easily secured than the summer comfort. To combine the two the following plan is recommended: Build ten feet wide and twenty feet long with seven foot posts for sides; have gabled roof, the ridge of which is ten feet from ground. Use strips 1x3 inches two inches apart, to cover the east side. Board up the south side three feet from the bottom, cover four feet with strips as on east side, then board up to the peak. Board up the north and west sides tight with inch boards. This will give a house warm enough for Mississippi in winter and a cool one in summer. The house is large enough for forty fowls, either kept as one flock or divided by partitions of wire netting. We would not recommend lengthening for a greater number as in other houses mentioned. The short length allows a greater circulation of air than a larger house would, and air is needed for comfort in the South. In the South, or even at the North, for summer roosting-places open sheds may be built simply to protect from rain and draft—a roof and two sides, or three sides if the open one be toward the south and the highest. Build as an ordinary horse shed at the country meeting-house. We believe in fresh air in the summer time for comfort, but we are inclined to think that comfort and health both are better conserved (preserved if you prefer) in winter by having as little air admitted as possible, provided the house and fixtures are kept clean and sweet. We would take our chances without the ventilation anyway. If any ventilation at all, we would have the air come in some round-about way and go out by a box ventilator that came within two or three inches of the floor on which the droppings fell.

BRICK OR STONE POULTRY HOUSES.

When brick or stone are cheaper than lumber, or when they can be afforded, they may be used with equal advantage to lumber and will make warm poultry houses. If ceiled up with lumber, or furred out and plastered and sheathed up with tarred paper, they would be ideal poultry houses for warmth and dryness. The objection some raise to plastering a poultry house, that the fowls destroy the plastering for the lime in it, ought not to hold. Enough lime should be furnished the poultry to prevent this. Poultry houses built of timber may be plastered if desired. These are easily kept free from vermin, and with an occasional whitewashing they are kept sweet and clean.

LOG AND SOD POULTRY HOUSES.

Good warm and dry poultry houses may be built of logs or sod and in banks where drainage is good. Have them face the south and provide light and warmth by glass in the front. No matter what the material of which the houses are built either have double-sashed windows or shutters that may be closed over the windows winter nights.

A PRACTICAL POULTRY HOUSE.

In the chapter, "Chickens on the Farm," we suggested that a shed attached to the poultry house was often convenient and agreeable to the fowls. Such a shed may be provided for the outdoor exercise of poultry in winter much cheaper than the same room (space) made frost-proof can. In other words, the same number of fowls can be warmly and healthfully housed in winter at less expense where the housing consists of a frost-proof roosting and nesting place and a shed, than where the whole room is made frost-proof.

Fig. 12 represents such a house and shed combined. This, like most of the houses we illustrate, may be extended to any desired length. If for but one flock, we should have the shed at the east of the house, which will give it a sheltered location and not shut off the light from the hen-house. The shed may be closed with lath or wire netting if it is desired to confine the fowls temporarily at any time. A few perches in the shed would make it an agreeable, healthful roosting place in warm weather, provided the droppings were removed as regularly and the same care taken to preserve cleanliness as in the house.

A WELL LIGHTED POULTRY HOUSE.

As stated at the opening of this chapter, there is some question about the benefit of a large part of the south front

FIG. 12—A PRACTICAL POULTRY HOUSE.

of the poultry house being made of glass. One objection to so much glass is the bright glare produced by so much light through glass. From Poultry Topics we get Fig. 13, representing a well lighted poultry house which has a good amount of

FIG. 13.

light and warmth without so much glare, as there is considerable part of the house that will be shaded by the sloping front. The low part covered with glass will make a fine place in which the fowls can sun themselves on severely cold days.

Upon days that are not so cold this glass will admit just as much warmth and light as the same surface of glass would higher up, without producing so much glare and direct heat out of which the fowls can not get, should they wish to. One other advantage this low, nearly flat, mode of admitting sunlight is, the night shutters for protection against cold are more easily adjusted than where the windows are larger and more nearly perpendicular.

AN OUTSIDE INCUBATOR CELLAR.

As suggested in the chapter on "Incubators and Brooders," there is no place for running an incubator where the

FIG. 14—OUTSIDE INCUBATOR CELLAR.

temperature can be more easily kept even than in a cellar and one specially constructed for the purpose will be the more satisfactory. This need not necessarily be under the dwelling house, barn, or brooder house; one may be made independent of any building. Fig. 14, which is reproduced from the American Agriculturist, shows an outside incubator cellar upon a broiler farm in Hammonton, N. J. This arrangement gives perfect satisfaction, therefore it may be considered the best.

HEN-HOUSE OF M. K. BOYER.

Perhaps there is no better posted broilerman in the country than M. K. Boyer, Hammonton, N. J. From his "All about Broilers" we reproduce the illustration Fig. 15, which shows the elevation and ground plan of his hen-house. It is thirty feet long, twelve feet wide, and nine feet high to the eaves. It is covered with Neponset lining paper. In the ground plan A shows door to entry. B—Entry. C—Door

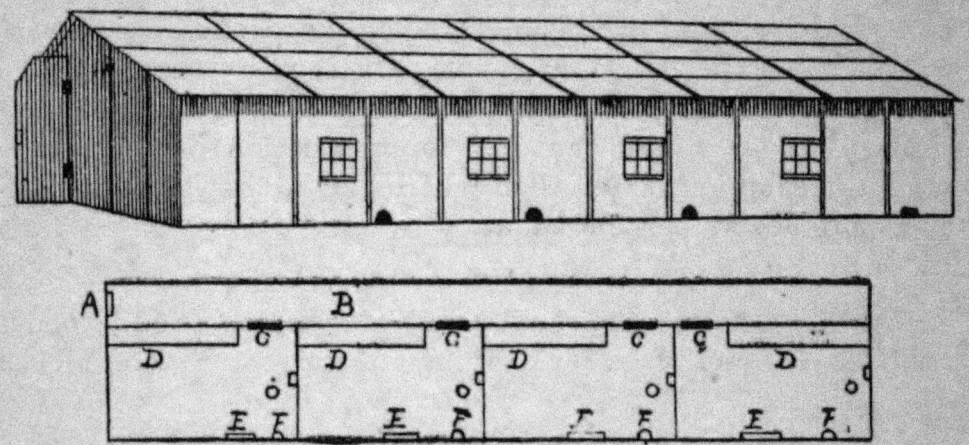

FIG. 15—ELEVATION AND GROUND PLAN OF M. K. BOYER'S HEN-HOUSE.

from entry into pen. D—Roosts, two feet high with platform under to catch droppings. E—Windows. F—Small doors for fowls to get out into yards.

FIG. 16—A GLASS-FRONT POULTRY HOUSE.

The pens are 7½x9 feet, the entry three feet wide. Nests are arranged under the dropping boards, with door opening into entry, making it convenient to gather eggs. There are drinking fountains and oyster-shell boxes on the right of each pen, as shown in diagram.

A GLASS-FRONT POULTRY HOUSE

Our illustration, Fig. 16, shows a neat little poultry house with roof sloping back nearly to the ground, while the front is nearly all glass, with board shutters to cover it when necessary. The windows are protected inside by wire netting.

A very satisfactory house in which 100 hens were wintered, was forty-eight feet long, eight feet wide, seven and one-half feet high in front, and four and one-half feet in rear. Shed roof shingled. South side and east end double boarded, and banked up to the bottom of the windows, leaving a passageway to the door in east end, and also one for the fowls. On the north side and west end it was banked to the eaves. Four windows in the south side, with board shutters. Two box ventilators extended up through the roof. There were perches enough put up along the north side and west end to accommodate the fowls; under these was a wide platform; under the platform were the nest boxes. The house was divided into four rooms by means of three lath partitions, each partition having a lath door.

A STRAW POULTRY HOUSE

was built thus: A stout framework was made of posts and poles, and then the straw was stacked several feet deep all over and around it, leaving only the south side open. Rough boards were used to partition off a roosting and laying room at the back side of the immense shed. There was not a pane of glass in this "poultry house," and the only cash outlay was for the few boards, nails and spikes; but it was a comfortable place for poultry, and the hens that were wintered there laid right along regardless of the outside weather, while the next neighbor who burned his straw stack to "get it out of the way," and let his fowls roost anywhere they could find a place because he "couldn't afford to build a poultry house," bought eggs for home consumption.

FLOORS IN POULTRY HOUSES.

Whether to have a floor to the poultry house is a question which every poultry-keeper must decide for himself. If the ground location of the poultry house is dry and well drained any other floor than the soil is unnecessary; where there is any likelihood of dampness it is better to have either a plank or cement floor. There should be an inch or more of dry soil or ashes on the floor; renew this as it becomes soiled. If six inches are put on at once it can be raked off once a week or so and kept clean.

INSIDE FIXTURES.

As stated in our introduction to this chapter we are not advocates for many fixtures. The nests we prefer, where

FIG. 17.

the laying-room and the sitting-room are connected, are those having two fronts, or interchangeable fronts. With these, all that is necessary, when a hen wishes to sit, is to reverse the back and she comes off into a room where she can feed with no danger of the laying hens taking possession of her nest during her absence. Where no hens are set, a simple box or row of boxes is all that is required. Have enough roosts and nests—enough to accommodate the number of fowls kept in each pen. If the perches are more than two feet from the ground floor have some form of ladder by which the fowls can reach them. The form easiest made is that shown by Fig. 17. A few cleats or bits of lath nailed to a six-inch board form it. A more substantial and elaborate one is shown in Fig. 18, and is better than the other for heavy fowls.

Where there is a passageway running lengthwise of the house, the nests, may be arranged so that the eggs may be gathered from it, and the necessity of entering the laying room be avoided. The nests and other fixtures may be under a second floor, over which are the perches, as shown in Fig. 11.

We consider flat or half-round perches the best. They are all the easier for the fowls if in the rough state, that is, as

FIG. 18.

sawed from the tree, or with the bark on if the half of a pole. Have all perches on the same level, or there will be continual striving after the higher and crowding there, while the lower ones will be deserted.

There may be those who can more readily obtain kegs than boxes. Such can make capital nests from them as suggested by Fig. 19. Simply cut a hole in the side near the end that has not been knocked out when opening the keg. If the breed that is to use it is a heavy one, put a block of wood or other "stepping-stone" in front of the keg. One objection to keg nests is the fact that having so many cracks or seams, they present more harboring places for lice and other vermin than boxes. Whether boxes or kegs are used, have the nesting material as near the upper edge of the nest as practicable, without danger of the eggs rolling out. This will prevent the hens from jumping upon and breaking the eggs that may be in the nest as they approach it. Have the nests near the floor or else approaches to them, for the heavier breeds very much dislike to fly to a

FIG. 19.

nest. Some authorities say they will not get into a nest that requires them to fly a foot. Have all nests, of whatever or however made, movable, that they may be easily taken out, cleaned and fumigated.

FEEDING AND DRINKING UTENSILS.

Feeding troughs or boxes, and drinking vessels of some kind are a necessity in every well regulated poultry house.

FIG. 20.

The best are very simple and cheap. Fig 20 shows a good-looking, serviceable trough for either food or water—one that can be cheaply and quickly made. It can also be easily cleaned and scalded. If dressed lumber is used in its construction all the better. The only objection to this trough is that fowls will sometimes forget their "manners" and put their feet in it; but this can be overcome by hinging on a slat cover, or by making it like Fig. 21.

Figs. 22 and 23 show an Iowa feeding or drinking trough. The inventor gives the following directions for making: "Take a board ten inches wide of any desired length (I pre. fer six to eight feet); on each side nail a three-inch strip

FIG. 21.

same length, leaving the board full and the trough eight inches clear inside; then take two pieces ten inches wide and eleven inches long, draw a mark across them six inches from one end and five inches from the other, nail on each end so as to have five inches above trough and two below for legs, then take a two-inch strip same length as inside of trough

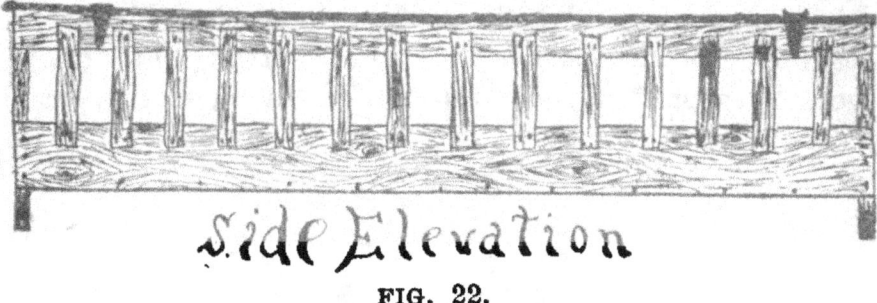

FIG. 22.

and nail on each side inside even with top of end boards, leaving an open space of three inches on each side for chick-ens to reach into trough and eat. Take a board ten inches wide and length of trough and place on the top, securing it in place by putting leather hinges (or others if able) at each end on one side so as to lift up cover to put in feed, and you have a rack the fowls can feed out of without soiling their feed." In the illustration our artist added the upright slats; made in this way, it will keep the little chickens out also.

If boards from which to construct troughs like some of those mentioned above are not at hand, a good receptacle for food and drink can be made by taking a nail-keg, driving down the hoops and securing them firmly with nails; then

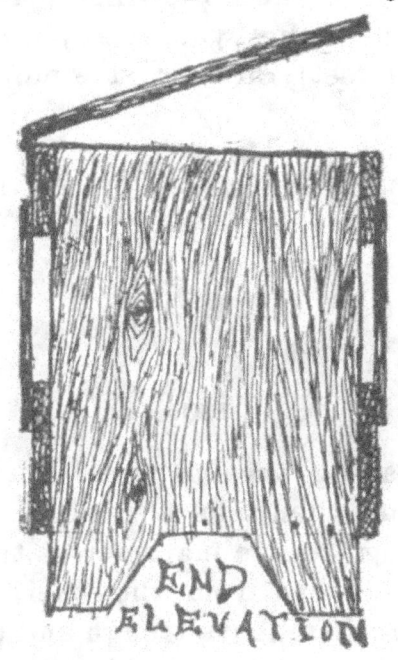

END ELEVATION

FIG. 23.

with an auger and knife remove a section of every other stave four or five inches from the bottom. In this "open-work" keg put the food, and while the hens can get at the food easily, they cannot get in and waste it by tramping

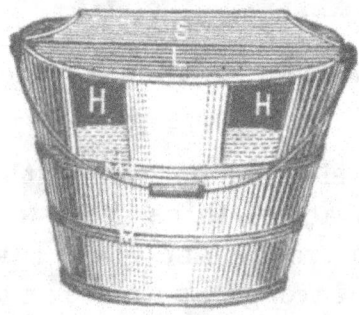

FIG. 24.

around in it. These kegs can be used for watering fowls by placing a dish of water inside, or a common wooden pail can be fixed for a convenient drinking vessel, as shown in Fig. 24.

Our hens have to be content with an old flat-bottomed iron kettle, old iron pans and gallon stone jars. These are all easily cleaned and kept sweet, and the jars neither rust

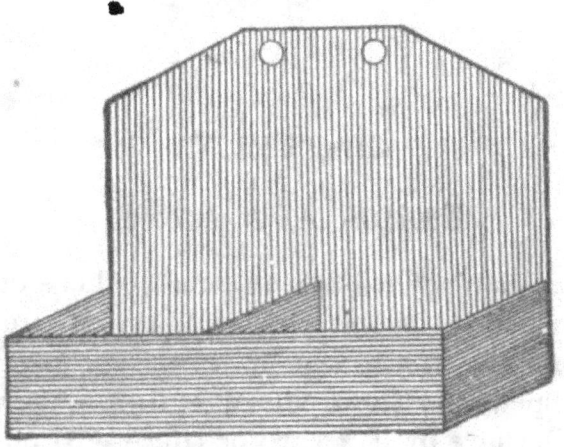

FIG. 25.

nor decay. For flocks of small chicks there are no better drinking plates than the saucers of earthen flower pots.

Besides the feed and water vessels, the only furniture needed is something to hold a supply of crushed oyster shells, gravel, charcoal, and other grit. For this purpose any common box will do, placed anywhere on the floor. If you wish something handy to remove from the building at cleaning up time and also out of the way while in use, make a box like the one shown in Fig. 25, and hang from spikes driven into the side of the building.

CHAPTER X.

YARDS AND COOPS.

With some poultry raisers, especially those on farms who have small flocks, all in the way of keeping arrangements is supplied when the house and fixtures are provided. With others, and especially those who dwell on a "town lot," the first thing to be sure of is a yard or there will most surely be unpleasant feelings among neighbors. It is our opinion that no one thing has been the cause of so much unkindly feeling between neighbors in small towns as "those hens." Again, the villager wishes a garden on his 50x125 foot lot, and he should have it, but hens and a garden never did well on the same piece of ground at the same time. With these two factors, peace in the neighborhood and a successful garden, entering into the poultry business, there must be another – some way of confining the fowls.

In these latter days when wire netting is so readily obtained, many will prefer using it to any other material. All that is necessary to build a fence of this is to have posts of the desired length firmly set in the ground and the netting stretched tightly from one to the next and stapled there. The distance the posts should be apart must be regulated by the height of the fence. In a fence eight feet high we would not advise having the posts more than eight feet apart, six would be better. In a fence four feet high the posts may be ten or even twelve feet apart if they are firmly set, and the netting stretched tight and securely fastened to the posts.

The height of the fence, whatever the material used, must depend upon the breed of fowls to be confined.

A fence four feet high will keep in the heavy breeds. It will need one eight feet high to keep in the light breeds, and

given a wide space to "get a start" we doubt if even eight
feet would be high enough to confine some birds.

Next to wire netting, dressed pickets, two inches wide
and an inch thick, make the neatest yard. Perhaps, in the
eyes of some, a fence made of such material is even neater
than netting. If painted they do make a very pretty fence.
The vast majority of poultry keepers will not use paint.

Taking the country at large there is no doubt but that
the ordinary lath which may be had at any lumber yard will

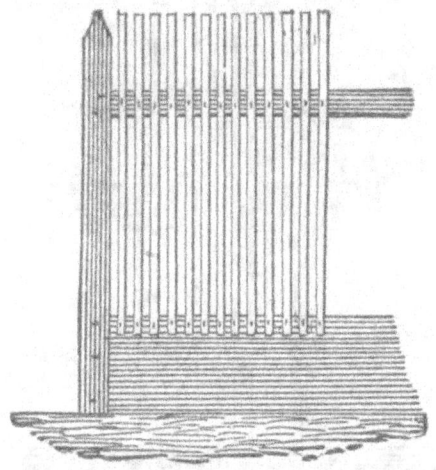

FIG. 1.

be the material most used. For a low, substantial fence
made of lath, the mode of construction shown in Fig. 1 is a
good one. In this the bottom board is a foot wide, giving a
fence nearly five feet high. If a fence four feet high is
sufficient it may be made by using 1x4 strips as the framework; in this case the bottom one may be six inches or so
from the ground.

Where a fence eight feet high is required a good one may
be made of lath or two-inch pickets if constructed as shown
in Fig. 2.

If the run or yard is for a small flock it will take no
more lath and not as much work to make a covered run. We
have made several in this way: Take strips 1x4 and nail
lath to them. For the sides nail the lengths to posts set firmly in the ground; for the top lay the lengths on crosspieces
laid on top of the posts to which the side lengths are nailed.

If the run is only four feet wide you have used a third less lath than to build a fence eight feet high. If it is eight feet wide you have used no more, even if the laths are as close together in the top as in the sides, which is not necessary. Such a yard will stand much more wind than one that is surrounded by a fence eight feet high. To move it cut the nails that hold the side lengths to the posts, take up the

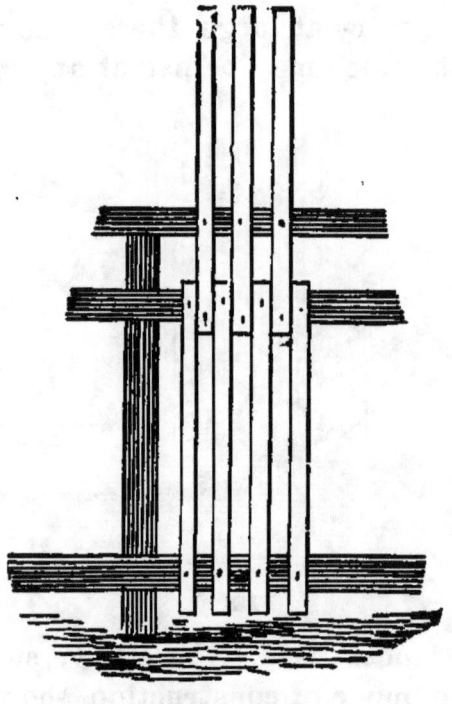

FIG. 2.

posts and reset; two men can easily carry the lengths. The only "quite a job" about it is taking up the posts.

When a fence that can be easily moved is required, make one without posts, as shown in Fig. 3, or have a wide board at the bottom, and nail the lath directly on that. This makes a fence about five feet high, and one that is difficult for the fowls to fly over. Waldo F. Brown, the inventor of the fence, enumerates its advantages as follows: 1. "Its cheapness, material costing about fifty-five cents a rod. 2. Nearly all the work of making it can be done under cover in stormy weather, and the fence can be set up when the ground is frozen so hard that it would be impossible to dig

post holes. 3. It can be easily moved from one place to another.''

To prevent this fence from blowing over in a gale, Mr.
Brown says: ''Drive a short stake at each pair of trusses

FIG. 3.

and drive a nail through the brace-board into it. Most of
these stakes need not be more than a foot above ground
when driven, but occasionally a stake should come up to the
top of the truss.'' The panels of this fence are eight feet
long, the trusses three and one-half feet high.

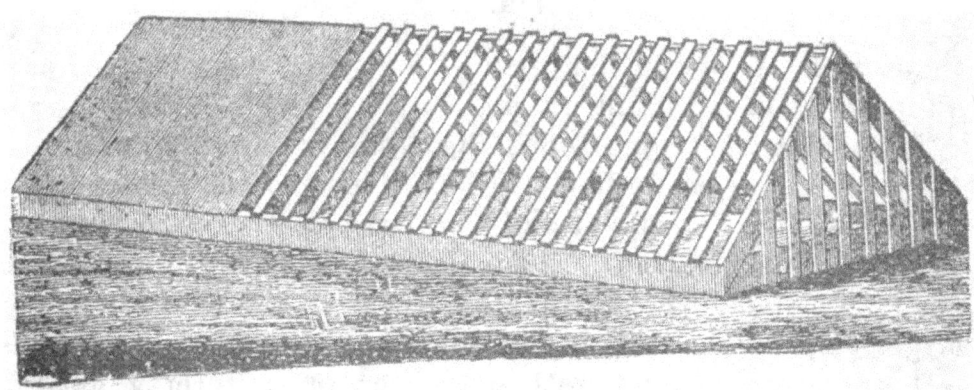

FIG. 4.

If only half a dozen fowls are kept, or a small number of
Bantams, they may be confined in a yard with house attached
as shown in Fig. 4. If the house part is built of half-inch
stuff the whole outfit may be easily handled by two men and
moved from place to place on the lot so as to give the fowls
a fresh run every day. We have such a yard in which we

shut up broody hens when we do not wish them to sit. Put
the yard where the inmates can see the balance of the flock
eating or enjoying themselves scratching in the straw and it
takes them but a short time to make up their minds they do
not want to sit; we think a much less time than it used to
those we in former times shut up in dark places. We give no
feed but plenty of water to the shut-ins. The third day we let
them out to feed with the rest. If they return to their nests,

FIG. 5.

they are again put in the prison. This yard may be used as
a feeding coop for young chicks.

Speaking of feeding coops, Fig. 5 shows one we have used
for years. It is easily made of lath, four short corner pieces
2x2 or heavier if desired, and two crosspieces 1x2 to which
the top lath are nailed. In such a coop or yard, little chickens may be fed with no interference from the old fowls or
larger chickens. There is one objection to this coop. If the
lath are near enough together to keep out a "pretty good-
sized chicken" the little ones are soon so big that they have to
struggle like little heroes to get through the spaces, and we
have an idea that more than one chick with a deformed
breastbone was given it in this struggle. Because of this we
prefer the coop constructed as shown in Fig. 6. It takes a
trifle more material and a little more time to make, but
when once made it is good for years, and enough better, we

think, to pay for this little extra outlay. These coops, also, are handy for shutting up broody hens or keeping a hen inclosed while her chicks run about.

CHICKEN COOPS.

Unless you are engaged in the broiler business only, you will need some coops in which to house the chicks after they are taken from their nests. The coop easiest made and soonest prepared for occupancy is an old barrel laid upon its side and a few stakes driven in front of the open end. That will

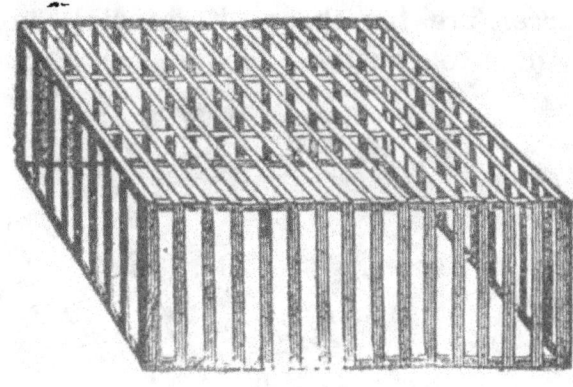

FIG. 6.

do in an emergeney or for a few days, but for a continual abiding place either for the hen or the chicks it is not desirable. It soon becomes foul and is not easily cleaned.

To be the most successful, even in the chicken business, one must take time by the forelock and keep his work ahead of him. Getting the coops ready during the winter before they are needed is one way in which work may be kept well in hand. The shape of the coop is a matter of little importance; its looks are nothing to the hen or chickens, but on the lawn or grass plot a neat coop is attractive and adds a look of thrift to the place not given by a ramble-shackle affair. One thing besides neatness, before it, for that matter, that must be secured in building coops, is proper size. Make them large enough that the hen may be comfortable while making it her home. Have the coops high and roomy. Four square feet on the ground and high enough iu the middle so the hen will not bump her head when standing upright is small enough to be called "roomy."

All coops should be placed on high, dry ground, or have floors. These may be movable or fastened to the coop, as the owner wishes, but for ease in cleaning, the movable floor is much preferable. Coops without bottoms should not remain long in a place. If for any reason they cannot be moved, the ground under them should be frequently scraped, and a little fresh soil or sand spread over the surface after each scraping. When the bottoms of the coops are scraped, whether movable or stationary, they should be covered with an inch or so of fresh soil or sand.

Where barrels are the cheapest material that can be obtained from which to construct coops, very convenient ones

FIG. 7.

may be made, as shown in Fig. 7. Such a coop should not be made of a common flour barrel, simply because a barrel of that size is not large enough. Use a hogshead, lard tierce, or a kerosene barrel. Here are the directions which the genius who invented this coop gave for making it:

"Nail every hoop on each side of a seam or line between the staves, with an inch nail; clinch nails are best; after nailing the hoops all tight, saw off the hoops on each side of the seam. (It is understood that if an oil barrel is used, the "saw" used must be a cold-chisel, and that to get the "inch nails" through the iron hoops a punch must be first used.) This leaves you with two half barrels, or half circles, each of which will make a fine coop. The pieces that formed the bottom (or top) of the barrel can be nailed in again at the back end of the coop; the upper part should be fastened with leather hinges, so as to open at pleasure; bore a few auger holes in the back for ventilation. Nail two parallel laths on the front, to fasten the slats to, make two of

the latter to slide in and out; make a floor of rough boards
to stand the barrel on, just a trifle smaller than the latter,
so that rain will be shed outside on the ground. A coat of
thick paint, or some waterproof roofing, tacked on, will
complete as nice a coop as any one need want, and at little
or no expense."

Where half-inch stuff costs no more than inch boards it
will be better, as far as practicable, to use it in the construction of chicken coops, as it makes them so much lighter

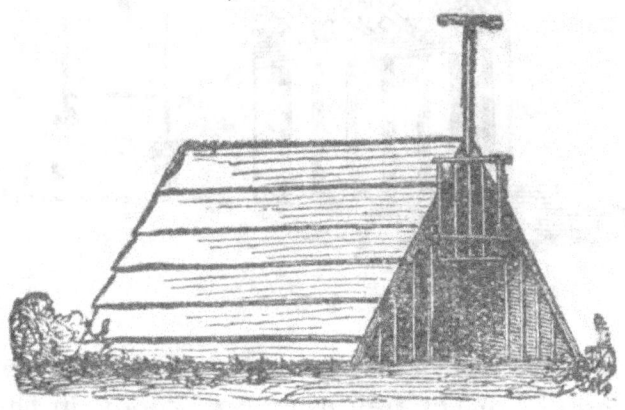

FIG. 8.

and handier to handle. Fig. 8 shows a convenient coop.
From it the hen can be caught and taken without the risk
of her dodging out under one corner as the edge of the coop
is raised to insert the hand to catch her. The sliding door
of the front can be easily raised with one hand while the
other takes biddy and removes her without danger of break.
ing her neck or back by a sudden "let go" of the coop should
you notice her making way for liberty, while you are trying to catch her in one that had no such arrangement.

Fig. 9 shows a coop in which beauty and usefulness are
combined. The upper half of the front is of wire netting,
which admits air and light, while the projecting roof keeps
out sun and rain. The lower half is made of perpendicular
slats, placed far enough apart to allow the chicks to pass in
and out. Hinged to the bottom of coop in front is a door,
A, which can be turned up over the slat front and fastened
with a wooden "button," or with a hook, making all secure

at night. This coop also has a door, shown at right side of
cut, through which the hen may enter or be taken.

It frequently happens that a hen will not allow chicks
from other broods to associate with or visit her brood, and

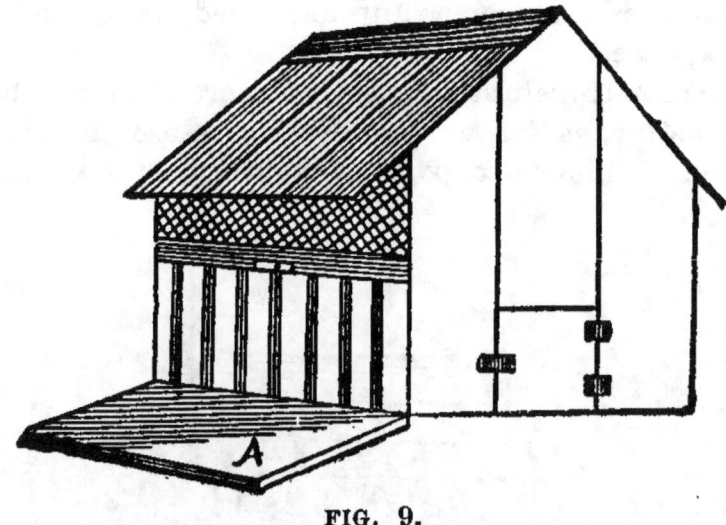

FIG. 9.

will punish with death all intruders on her territory. In these
cases the chicks of the neighboring broods on the lawn or in
the lot must be kept from danger by surrounding the coop of
the beligerent hen with wire netting or an inclosure of lath

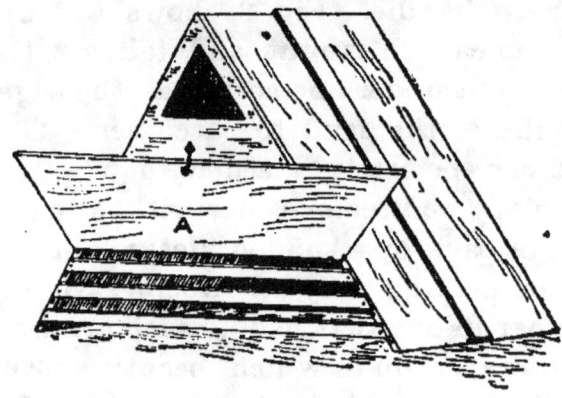

FIG. 10.

or other material through which her chicks can not go out or
neighboring chicks come in.

Taken all in all, there is probably no more popular chick-
en coop than the old-fashioned triangular one shown in Fig.
10. It is easily made and takes but little lumber. With so

few slats on as shown in the illustration it is rather uncomfortable.for the larger hens, as they must stoop some to get a full view of the outside world. The drop door, A, makes it easy to protect the brood at night, and except in very warm weather the air received through the screen-covered square in the gable will afford sufficient ventilation.

We heartily endorse the movable floor, if there must be a floor in coops. Fig. 11 represents a good one that is easily made. Take inch boards, A A, and nail, dressed side up, to the cleats C C. Let both ends of both cleats project three inches,

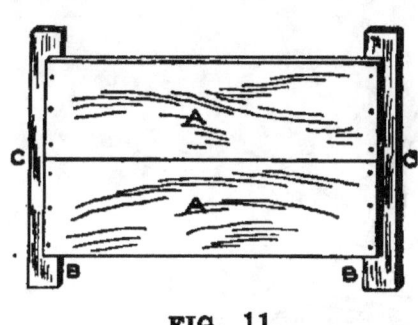

FIG. 11.

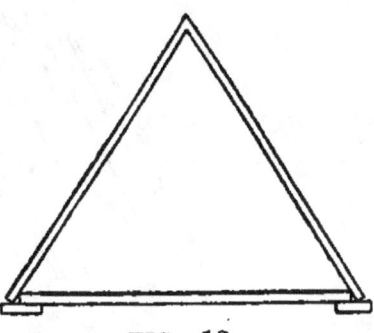

FIG. 12.

and the outside edge of each two inches. Make the floor just the right size that the coop will rest entirely upon the ends and edges of the cleats, so that when the door A, shown in Fig. 10, is closed all rain will be shed outside the floor. When the coop door is closed, fasten it by pegs stuck in the holes bored in the cleats at B B, Fig. 11.

We know of no better way of keeping rats, other marauders and wet out of a chicken coop which is as cheap and simple as this. During cold days or nights it may be closed in a few seconds so as to keep out the cold as much as possible, while allowing sufficient air. If made of half-inch stuff these coops are very easily handled.

Fig. 12 shows a section of a triangular coop with the movable floor in place.

Once a week or oftener, when about to clean the floor, place a smooth piece of board the thickness of the cleats opposite one end of each and slide the coop slowly lengthwise of the cleats away from the floor, which may then be scraped with a hoe. Then put on a layer of earth and slide the coop back to its former place. The chickens are sure to get out of the coop, and so out of the way, just as soon as you move the coop a few inches, and if care is taken none will be hurt.

Fig. 13 shows another triangular coop. We prefer this to the one illustrated by Fig. 10, as the slats are up and down and the door, A, may be used as a feeding board when opened, as shown in the illustration. This coop is supplied with air

FIG. 13.

through the ventilator made by the opening covered with wire cloth. If there are no marauders—rats, cats, minks or skunks—about there will be no need of netting over this—no need of closing the door, A, only for warmth or as a protection in case of a driving storm. This coop to be without objection should have a small door in the back, through which the hen may enter or be caught when desired.

NUMBER OF CHICKENS TO EACH HEN.

Having roomy coops in readiness for biddy when she comes off with her brood, there may be a temptation to give her more than she can successfully hover and rear should there be two hatching out at the same time. Do not yield to the temptation. Because she is willing to adopt some extra chickens, do not give her all that she can possibly stretch her wings over; from fifteen to twenty, according to the size of the hen and the season, will do much better than a larger family.

If desirable to give the hen more chickens than she hatched, the proper time to do it is when you remove her from her nest to the coop before she has had an opportunity to count her family. If you slip the extra chicks under her just before she leaves the nest, or just after she settles down in the coop, she will never know but that she hatched them all. Occasionally a dark hen will object to mothering a single white chick, even though she hatched it, but if given a half-dozen or more white chicks she will mother them all. And there have been white hens that would not own a black chick, but such cranky hens are not common; generally a hen will mother all the chicks given her if the broods are doubled up at the right time.

When removing the hen and her brood to the coop is the time to use more lice preventives. If the nest was prepared, the hen dusted, and there has been a dust bath handy during incubation, the hen and chicks will probably come from the nest free from lice, yet both hen and chicks should be examined. If any lice are found dust the hen with insect powder before you put her into the coop; then when she gathers the chickens under her they will get their share from her feathers. Some recommend whether lice are found or not, putting a drop of kerosene on the head of each chick as a preventive against the big head lice. One writer says; "The quickest and best way to apply it is to have some in an old cup, and as you pick up each chick just dip the tip of the forefinger in the kerosene and rub it on the top of the chick's head, taking care not to get too much of the oil, and to keep it out of the eyes. Just a twitch is sufficient."

We are afraid of kerosene about little chicks; there is so much risk of getting on too much and killing the chicks as well as the lice, that we have never used it. Rather chance the old hen with plenty of dust and insect powder or snuff about her to keep the lice off. Keep the coop clean, oil in the cracks of it if you like, and give the hen and chicks an opportunity to take a dust bath when they feel like it, and the probabilities are you will lose less chicks from lice than you will from the use of kerosene. Others have.

CHAPTER XI.

BROILERS.

Broilers have made money for some men, they have lost it for others. The difference in results was occasioned by the difference in the men, their ways of getting the birds from shell to market, not in the broilers. Experience is the best teacher, but it is not advisable to pay too much for the teacher. It would be folly for a man, because he had money, to pay a professor of mathematics in some of our colleges $5,000 a year to teach his boy the rudiments of mathematics when he could get a teacher well versed in primary arithmetic for a few dollars per month. So it is folly for a man to go into the broiler business on a large scale for the experience. Learn by beginning in a small way. Read the experience of others; if possible visit the broiler establishments of those who have been successful in this line of poultry farming. But after reading and visiting do not think you know it all, and invest several hundred or a few thousand dollars and lose it all in learning how to make broilers pay. Better make broilers a side issue till you have experience, or make it winter business in connection with some work that is at its height in the summer—fruit growing, for instance.

A good plan for a start is to have a small brooder house and two incubators with a capacity of two hundred eggs each. With this little outfit you could run a small broiler business in connection with other work from which to get a living. As one becomes posted in the business it may be enlarged until the whole time is devoted to it. In that broiler center of this country, Hammonton, N. J., there are men who from October to July virtually abide in their brooder houses.

In going into this business remember there are some

things that have to be had whether it is run on a small scale or a large one. We refer now to broiler raising as a business— not the raising of a few chickens on the farm to be sold as broilers. First, there must be incubators and a place for them; second, a brooder house and brooders; third, eggs; and fourth, fuel, feed and other running expenses for four or five months before there is a cent income from all this outlay. If only an incubator or so is used the "place" need not be a house built expressly for the purpose. The same is true as regards the brooders, though a brooder house is more necessary than an incubator house when the business is run in a small way.

If a farmer wishes a small broiler plant in connection with his other farm operations, he might put it in charge of his wife or daughter, but he should see that all the cleaning up, carrying water and other such work is done by the men folks—that is not woman's part of the business. Some of the best and most successful incubator managers at Hammonton are women.

EGGS FOR BROILERS.

Among dairymen there is a difference of opinion as to whether a man who pushes everything for milk should be at the expense of time, milk and money to raise his cows, or whether he should crowd a cow for all there is in her till her usefulness is gone as a milker, then let her go and fill her place by purchase. Each system of replenishing the dairy has its advocates. So, in the broiler business, there are those who hold that the broiler raiser should keep hens to produce the eggs he wishes to use. One puts forth his ideas in these words: "The wise broiler raiser, and the one who makes the most out of it, raises his own eggs. Last year, those who bought up their eggs averaged a hatch of about thirty-five per cent. Such costs cut deep into the profits."

To produce broilers from eggs that are home products requires much more land than when the eggs are bought from those who keep hens for their eggs. The broiler raisers who buy all or a portion of the eggs they use in the production of broilers are just as earnest in advocating the rightness— profitableness—of their way as their opponents. Mr. R. G. White is a successful broiler raiser and he buys all the eggs he uses. He gives his reason for so doing in these words: "I do not have much faith in egg farms unless the hens can have a free range. I have not ground to keep enough fowls on such a plan. So I do the next best thing by buying up my eggs from small flocks, and from birds that have my personal supervision, that is, they are mated and composed of such bloods as will give good broilers."

The farmer who raises a few hundred broilers as a side issue can produce the eggs on his farm or buy a few of his next neighbor to fill the incubator.

The amount of time needed to be given to the work of raising broilers depends on the extent of the business. Those who turn out large numbers give their whole time to the work. Attention to incubators and brooders, preparing the feed, giving it and water to the chicks, with the other necessary work connected with the business, causes the day to slip away. Each day is a repetition of the previous one, till the season is over.

PROFIT IN BROILERS.

Before figuring on the profits of broilers, the cost of producing them must be considered. At the West, where grain is cheaper than in the East, the cost may be less than at Hammonton. After an interview with nearly all the broiler men in that section, the following estimate of cost is made. Eggs are reckoned at two cents apiece. To fill a 200-egg machine costs four dollars. To supply the heat with oil twenty-one days costs twenty-one cents. When the eggs are tested on the seventh day twenty-five per cent of them are thrown out. This leaves 150 eggs; if sixty-six per cent of these hatch, there are 100 chicks that have cost $4.21 including the oil. Allowing that twenty-five of them die, the seventy-five will have cost about five and a half cents apiece. To raise a chick up to a pound and a half or two pounds (broiler size) costs, including brooding, but not labor, about fifteen cents a pound. A two-pound broiler costs, then, thirty cents for raising, added to five and a half cents for the foundation, a total of thirty-five and a half cents, with nothing for labor. These figures are based on calculations made by M. K. Boyer. P. H. Jacobs says "the cost per pound of broiler for food only is five cents." The average price for broilers where these figures were made was thirty cents per pound, giving sixty cents for a bird that cost thirty-five and a half cents; but before it is sold it has to be dressed, which costs five cents, leaving nineteen and a half cents to pay for the labor. Where they are raised by thousands this figures up a nice little sum. Broilers sell for no such prices in the West this season, nor do Western broilers in the New York market. Sixteen cents per pound is the best quotation we have seen this winter ('93 and '94).

RULES FOR RUNNING BROODERS.

Having decided to go into the broiler business, decide upon the incubator and brooder to be used and when obtained follow the directions for running them as given by the manufacturers. There are a few rules that will hold good in the running of all brooders. 1. Do not have them too warm. 2. Keep them scrupulously clean. 3. Have sand in the run. 4. Keep free from dampness. 5. Do nót overcrowd. Following is the room space allowed by a very successful broiler raiser. The size of the "mother" (brooder) in his pens is three feet square, the brooder floor is four feet six inches, the yard inside the pen is four feet six inches by five feet; and the outside yard is four feet six inches by sixteen feet. He aims "to have not over 150 chicks in the brooder the first week, although 200 could be managed. When they run to three-fourths of a pound I thin down to 100. After that as a chance offers I take a few out, until only seventy-five are in by the time they reach the marketable size."

FEEDING BROILERS.

The one point aimed at in feeding broilers is forcing them to the market point in the shortest possible time. The food is selected and fed with this in view. The Hammonton broiler men have no fixed rule; they vary the diet. M. K. Boyer follows this system of feeding: He has rolled oats before them from the start, with boiled milk as a drink. After a few days he gives a mash made of two parts of bran, one part corn meal, and a handful of meat scraps, to a pail of the mixture. This is scalded several hours before feeding, and fed warm—not hot nor sloppy. When about two weeks old. cracked wheat and cracked corn are added to their diet. Gravel or grit of some kind is constantly before them. About twice a week he adds bone meal to the mixture. He does not feed boiled eggs nor any condition powder. Meal and bran give him fine chicks. He feeds them once in two or three hours the first two weeks of their lives. After that three times a day. As an appetizer he feeds roast potatoes, cut in half. When the chicks are about a month old he changes the feed by giving equal parts of ground corn, and oats and bran to which meat scraps are added. Cracked corn is kept before them most of the time. Mr. Boyer's testimony is that this course of feeding with suitable crosses or pure-breds gives marketable birds quicker than any other.

R. G. WHITE'S WAY OF FEEDING BROILERS.

Another successful broiler raiser of Hammonton is Mr. R. G. White. He has a system of feeding peculiar to himself. He makes a johnny-cake without vinegar and soda. He puts animal meal into the cake instead of prepared meat. He prefers the meal because it is finer. This johnny-cake is the feed his chicks get till they are a week old.

When they are a week old he drops the johnny-cake and feeds a mash composed of equal parts of corn meal, bran and middlings, with the usual amount of meat scraps. At this stage in the chick's life bowel trouble is apt to show itself. Ot this Mr. White keeps close watch, and regulates it with the middlings—lessening the quantity of middlings if the chicks become costive, and increasing the amount if they have looseness of the bowels. After trying this method for several years, he says he finds no trouble in keeping the chicks in the right condition. He feeds wheat and cracked corn only as a relish. Several times during the day, between meals, he throws several handfuls of grain to them, thus getting them to exercise in scratching and running about. Grit is constantly kept before them. He uses small, sharp gravel, and likes it better than oyster shells for the purpose. When the chicks are first put into the brooder a pan of ground charcoal is placed before them. If after several days they do not eat any of it, charcoal is put into their cake and afterwards into their mash, as Mr. White says they must have charcoal; in fact, he attributes much of his success in raising chickens to the use of charcoal. Every morning after the brooders are cleaned up oyster-shell lime is scattered on them. A pan of sifted coal ashes is kept constantly before the chicks. Besides affording a nice dust bath it gives fine picking ground for the young chicks.

MEAT AND DRINK FOR BROILERS.

Furnish the youngsters warm water to drink and give it to them in vessels from which they can drink without getting wet, for dampness is fatal.

A raw egg stirred in the feed of twenty-five chicks three times a week is beneficial. The eggs that were thrown out of the incubator at testing time may be used for this purpose. Hard boiled eggs are discarded by the best poultrymen as being conducive to bowel disease. If the chicks become chilled they are apt to be troubled with bowel disease. If it is more common with the chicks that are feathering out fast they should be given raw meat once a day, as the want of nitrogenous matter to produce feathers is the cause of the debility. For this reason Dorkings, Leghorns, Games and Houdans, which begin to feather earlier than Brahmas, Cochins, Plymouth Rocks and Wyandottes require more animal food than the latter, when very young.

RAISING BROILERS FOR CHICAGO.

There are no extensive broiler centers about Chicago, yet there are some broilers raised artificially for the Chicago market. The system employed by Geo. B. Stapp, Hope, Ind., presents some interesting features, and is described in the American Poultry Journal by Henry Lee. His building is seventy-two feet long and twelve feet wide, ordinary "drop siding" being used in its construction. A long furnace extends the full length of the building and is arranged to warm the different pens in such a manner as to make the ordinary brooder entirely unnecessary. This furnace is fired at one end. Along the top ordinary tiles are arranged side by side for warming the air, on the principle sometimes utilized in greenhouses. The furnace and the floors of the pens are covered with cement.

The incubator house, which is an auxiliary to this house, is built on exactly the same plan, but is only twenty feet long. The floors of both houses are covered with straw chaff to the depth of two inches. Ample ventilation is provided.

For the first three weeks of their lives the chicks are kept in the incubator house where the temperature is sixty-five to seventy degrees at six feet from the furnace. They are then removed to the brooder house, and placed in the room nearest the fire-box of the furnace, where the temperature is the warmest (65°), and at the end of each succeeding three weeks they are moved one room further away from the warm end of the building. The temperature in the eighth (last) room is five or six degrees lower than near the fire-box. Thus the chicks fresh from the incubator are given the warmest quarters, and by the time they have reached the further end of the brooder house they are of marketable size, and a succession of hatches is nicely accommodated. No other warmth is needed than that furnished by the long furnace. The chicks sleep upon the chaff without any covering. On a very cold morning they may be found scattered along the side of their room next the furnace, but as the temperature is uniform all along the furnace and the cement floor is also warm for some distance they never huddle together.

Mr. Stapp first tried using brooders in the rooms, but the chicks left them and went to the furnace—as he says "for comfort and fresh air."

It is a well known fact that a healthy young chick prefers to sleep with its head in the open air. If the fire should happen to die out at night, the furnace and floor would still retain sufficient warmth until morning. Thus it will be seen the great advantages of this system over the common brooder are freedom from crowding and the absolute security against loss in case the source of heat should from any cause be disarranged.

With these essential principles thus fully provided for,

any good system of feeding will make broiler rearing on a large scale a success. Mr. Stapp's bill of fare is: Boiled rice the first three days; rice and wheat cooked together for the next ten; after which, or as soon as they will eat it, come wheat and cracked corn. The first few days they are fed four to six times a day; after that, three times regularly. For variety they have the usual bran-and-meal mush seasoned with a little cayenne pepper; also meat, cooked or raw, three times a week. Their green food is chopped onions and a head of cabbage to pick at. Sloppy food is avoided and no water is given them until eight or ten days old. No sand is supplied until they begin to eat dry grain.

BREEDS FOR BROILERS.

It is pretty generally conceded that a common dunghill is not a profitable chick from which to make a broiler. A thoroughbred cock of breeds that produce good broilers mated to dunghill hens will give a better broiler than simple dunghills, yet these will not do as well as the first matings of pure-breds.

Of the pure-bred poultry that give excellent results as broilers we mention Wyandottes, Plymouth Rocks, Dominiques, Leghorns, though some prefer Leghorns crossed upon other breeds. Houdans have given good satisfaction; they are plump, quick-growing and their meat is of fine flavor. Houdans have been crossed upon Cochins, Brahmas and Wyandottes with good results.

The aim in crossing is to get a better broiler than either of those used in the cross would produce, hence we must use breeds that possess different merits to a marked degree. Use a male of a quick-growing breed and have hens with solid bodies and broad breasts. This gives meaty broilers.

The same result—broilers with meaty breast—is obtained by using Indian Game cocks on the large breeds. On this account, the large amount of breast meat, the Games are becoming a popular top cross. They are used on Dorkings and Langshans with good results.

A Wyandotte or Plymouth Rock cock crossed on Brahma or Cochin hens gives hardy chicks that grow fast and to a large size. Some claim the very best broilers are produced by

White Wyandotte cocks on Light Brahma hens.   White Leghorns on White Wyandottes produce very quickly maturing broilers.   One cannot go far amiss in procuring breast meat and plump carcasses if Dorking, Leghorn or Game cocks are crossed on hens of any of the larger breeds.   If yellow legs and skin as well as plump breasts are desired, use a Brown Leghorn cock and Brahma or Cochin hens.   Leghorns and Dorkings are also crossed.

DRESSING BROILERS.

Broilers are dry picked.   Their legs are tied to a rope suspended from above; the operator stabs the bird in the roof of the mouth with a killing-knife.   As soon as the bird is stabbed the operator begins plucking its feathers and before the bird is through struggling it is bare.   After this it is pin-feathered, then cooled off before packing for market. They should be shipped in neat boxes or barrels.

WHEN BROILERS SELL.

The earlier broilers are sent to market the smaller they may be.   Those weighing a pound or less may be sold in February.   The March market at the East demands birds weighing a pound and a quarter, while pound and a half birds are demanded for the April and May markets.   April and May generally give the best prices.   Early broilers sell best in New York City.   The prices in Boston, later, are generally equal to those in New York.   In Chicago pound and a half to two-pound birds sell in March.   It is well to sell Leghorn broilers when the market demands or will take light ones, for later their large combs are detrimental to their ready sale.

CHAPTER XII.

SCRAPS ABOUT POULTRY.

Most persons preserve scraps of some kind. Those who do not have them in a book, keep them loose in a box or filed away in some order. We have kept scraps about poultry for some time, and in the few remaining pages of this book we give our readers some of them.

A WHITE BRAHMA RECORD.

I began by buying a mother hen, a Buff Cochin; the chicks were White Brahmas—thirteen of them.

I bought, last summer, a Buff Cochin cockerel, not fully grown. I weighed him to-day and he weighed nearly thirteen pounds.

Last year from sixteen hens I had 129 dozen eggs. I raised forty-four chicks, seventeen of which I killed; weight of fowls killed, from four to ten pounds each.

My hens lay every month in the year. No month I had less than five dozen. My chicks that hatched in April began laying in November.

LANGSHANS ON THE FARM.

From forty Langshan hens I sold in one year 270 dozen of eggs. I sold them at the market price at a country store, and they brought me $40. We had, besides, all that we wanted to use in a family of four, and used about forty dozen for setting.

My fowls had to roost outdoors on the fences during the months of September and October, as I did not have my coops built for them. Had they been properly sheltered I would have had without doubt twenty-five dozen more eggs. My hens have been laying all winter, have had from eight to eleven eggs a day.

As I live on a farm I have not fed my fowls at all until it

froze up and snow came, so what they got would have been a total loss if the fowls had not been there. I am now feeding fifty-five chickens about six quarts of oats, wheat screenings and barley mixed. I pour boiling water on it the night before feeding, and give it to them as soon as they are up in the morning. At night I give them about one dozen ears of corn to shell for themselves. I keep fresh water before them every day. When the days are warm I let them out in the middle of the day.

EGGS AT LESS THAN A CENT EACH.

Following is the record of fifteen Plymouth Rock hens belonging to Mr. Wm. Feeley, Hull, Canada:

January...........................214
February.........................144
March............................283
April............................280
May..............................240
June.............................211 One hen killed
July.............................197
August...........................187
September........................130
October.......................... 75 ⎫
November......................... 4 ⎬ Hens moulting.
December......................... 37 ⎭
                                 ———
                                 2,002

The total cost of feed for the year, $17.90. Four of the hens raised thirty-nine chickens. The cost of the chickens' feed is included in above. One of the pullets commenced to lay when five months and three days old.

A FAITHFUL SITTER.

One poultry raiser writes, "I have a hen—an old brown one—that gives me more satisfaction as a sitter and mother than any other hen I ever saw. She must be four years old, but it seems to me the older she gets the better. She is a "scrub," a "dunghill," that's the only objection I have to her. Since the first day of March she has hatched out four broods, and has raised in all fifty chickens and has not lost a single one. It seems as though she has a way of her own, and I must say, it has proved very satisfactory this year, and if I am not mistaken she did equally well last year."

HOW TO RAISE TURKEYS.

Desiring to raise as many as possible to the number of hens kept for breeding purposes, I set the first laying of eggs the turkey hens lay, under the chicken hens. When they hatch I examine to see if they have any vermin on them; if so, I dust them with insect powder and keep them cooped closely for three or four days, until they get used to the call of the hen. If allowed to run out they are likely to stray after any hen that comes along.

I feed often, and very sparingly the first week with hard boiled eggs and corn bread crumbs, mixed fine and dampened with a little barley meal, onion tops and lettuce chopped fine. Milk should be given them as a drink, as it keeps them in a healthy condition. Corn meal dough should not be given them, as they are liable to diarrhea and it increases the tendency. Their food should be strictly fresh. Keep cooped in the morning till the dew is off the grass, until they are six weeks old; for cold spring rains and dew are fatal to young turkeys.

The second laying of eggs I let the turkey hen sit on and raise the brood. I do not pay much attention to them except to feed a little each evening to get them accustumed to come up at night and keep them growing. They will pick up most of their living in their rambles.—J. Ellars in Ploughman and Farmer.

GIRLS AS POULTRY-KEEPERS.

Take care of the hens and they will take care of you. Perhaps some of you will think this is not true, but if you will try it and attend to your business in the proper manner you will find that I am correct.

I speak from eight years of actual experience in raising chickens for eggs. I seldom see any statements or estimates from the country. I know there are a great many girls who, if they really knew that they could support themselves and lay up a little money every year, would go into the chicken business. If you live near enough to a city to make it an object to raise broilers you have an advantage that I do not, as I live too far from a large city to raise broilers, so I content myself in the egg-producing business. Your hen-house needs to be warm and good. but not expensive. I do not keep my chickens shut up any more than I can help. The most of the year they roam at will, gathering the waste from field and barn. A prominent chicken breeder called the other day to see how I kept the lice out of my chicken house. I told him I was raising chickens not lice. Keep your house clean, do not be afraid of a sprinkle of lime or ashes, the hens like it. Cleanliness is death to all nits.

I have never had any disease among my flock. An ounce of prevention is worth a pound of cure when it comes to cholera and roup, which are easily prevented but hard to cure.

I have tried a good many different breeds, but find the White and Brown 'Leghorn to be the most profitable for eggs. I raise my own chicks, starting early in the spring, keep all the pullets and dispose of nearly all the cockerels, always selecting the finest birds for breeding purposes.

I feed corn for supper every night the year round, wheat, buckwheat, oats and sunflower seed alternate for breakfast, always keeping pure clean water at hand. I give the chickens both sweet and sour milk, but use no pepper or stimulants of any kind; good feed makes good blood and that furnishes stimulus enough. I do not recommend slop feed of any kind, neither do I use it.

I ship my eggs to a regular egg dealer who furnishes me with egg cases that hold thirty-six dozen each.—Country Cousin in FARM, FIELD AND STOCKMAN.

HOW BUFF WYANDOTTES ARE PRODUCED.

Buff Wyandottes, it is publicly claimed, have been originated in at least three different ways: 1. By crossing the Golden Wyandotte and the Buff Cochin. The Golden Wyandotte itself originated in several ways—by a cross of the Silver Wyandotte and the Winnebago, the Silver Wyandotte and the Buff Cochin, the Silver Wyandotte and the Partridge Cochin, and probably also the Silver Wyandotte and the Black-Breasted Red Game. There is, therefore, inasmuch as the Silver Wyandotte had a portion of Asiatic blood, a considerable percentage of the blood of fowls with feathered shanks in this strain of the Buff Wyandotte, and it need not be surprising to find many chickens with feathered shanks for some years to come.

2. By crossing the Golden Wyandotte and the Rhode Island Red. Just what the Rhode Island Red has been produced from, no one seems to know. It is a fowl of about the size of the Wyandotte, varying in comb but usually single, and of a reddish-buff color. It has never been bred to standard requirements but has developed on the farms of Rhode Island and vicinity with but the one thought of producing a good farm fowl. It is believed to have some Cochin blood, and some blood of other breeds, in its make-up. The shanks are, for the most part, free from feathers. The Buff Wyandottes from this cross are scarcely as large as those from the Cochin cross, but are, as a rule, better layers and have fewer chickens with feathers on the shanks.

3. By breeding from Golden Wyandottes only, selecting each year those which showed the least black markings. This claim has been made by one or more Western breeders, and it is possible that a strain of Buffs might in this way be produced.

In addition to these three methods the writer knows of one gentleman who has produced Buff Wyandottes by using Golden Wyandottes deficient in black markings, White Wyandottes and an Argonaut cock. These three varieties – all clean-limbed birds—have given him some very good Buff Wyandottes which are breeding very well.

But every purchaser of a new variety should not expect to obtain the uniformity in chickens that can be had from long established varieties. It took quite a number of years to get rid of the feathers on the shanks of the Silver and the White and the Golden Wyandottes, and even now in the last named variety it would not surprise me to learn that some specimens show down at least between the toes. And the Buff Wyandotte, being of so much more recent origin, will probably require several years more of breeding before the results will be altogether satisfactory. It is, however, improving from year to year, and bids fair to become an established variety.

I think the action of the American Poultry Association in recognizing the Buff Wyandotte was somewhat premature, but it may hasten the perfecting of the variety. Time will tell. Purchasers, however, who expect every chicken reared from Buff Wyandotte eggs in 1894 to be true to standard description will assuredly be disappointed. I think having "no two alike" in a brood need not be the experience of all if the best breeders are patronized, but all the chickens hatched will not be likely to be true to type.—H. S. Babcock, in Country Gentleman.

GROUND BONE AND OYSTER SHELLS.

Do not neglect to supply sufficient raw bone, either crushed or in the form of meal. It contains lime, as do oyster shells, and it contains animal matter which is of great value. Bone when burnt is of comparatively little value over oyster shells, but when crushed or ground raw it has a value peculiar to itself. All classes of poultry are extremely fond of it. Care should be taken to have it pure and sweet. It is good for all classes and ages of poultry. For young chicks it should be used in the form of meal, mixing a small quantity two or three times a week with their soft feed, say a quart to a bushel of corn meal.

For young turkeys it is almost indispensable to prevent leg weakness. At about the time of their "shooting the red," when their health becomes established and they grow fast, the development of their frames and legs require a more liberal supply of bone material than can be afforded by the usual articles of food. It is well to begin to mix a little bone meal with the feed of young turkeys, and from the time they are four weeks old it may be used freely.

When there is trouble from soft-shelled eggs it may be quickly remedied by a liberal use of ground raw bone and oyster shells.

Bone and shells may be fed to fowls from a narrow box nailed to the side of the coop; we prefer this method, as it is less wasteful than throwing them on the ground. To use bone meal beneficially it is necessary to mix it with the soft food.

To promote laying it is necessary to have it ground coarse.

TO BREAK AN EGG-EATING DOG

from eating eggs proceed as follows: "Divide a heaping teaspoonful of tartar emetic into eight or ten doses. Break off the end of an egg, empty a part of the contents and stir into the remainder left in the shell a dose of tartar emetic. Confine the dog in a room, or tie him, and give him the doctored egg. In an hour or two he will be trying to turn himself wrong side out. As soon as he is over the nausea give him a second egg and a third, if he will eat it. When he refuses to eat the egg, and lets it lie by him for several hours untouched, pry open his mouth and force the egg down his throat. Afterward you may trust him in your hen house. The object in tying the dog is to let him get nothing else to eat while he is under treatment or he may think it was the last thing eaten that made him so sick. The idea is to convince him that eggs no longer will lie on his stomach."

A CELLAR POULTRY-HOUSE.

A writer who has had practical experience, says that a barn-cellar, provided it be free from dampness, forms one of the very best of hen-coops, especially during the winter.

Poultry must be warm if they are to yield eggs in cold weather. Barn-cellars are usually quite warm and snug; at any rate a little attention will render them so. The south and east sides will need a window to every six or eight running feet. The north and west sides are best banked up to the sills of the building.

EGGS FOR SETTING.

All eggs selected for hatching should be of the fair, ordinary size and shape usually laid by the hen or pullet. Reject all small ones and all very large ones. They should also be firm and smooth in the shell. Those unusually long or differing in any way very much from the usual character of those laid by any particular bird should not be used. It if taken for granted that only those laid by the best stock, even if common farmyard fowls, will be saved for hatching.

THE BREEDS CLASSIFIED.

When Wright's Poultry Book, which for years was the

standard authority, was written the chief breeds of poultry were, for economic purposes, classified as follows, the order of naming representing as nearly as possible their average comparative value, though this varied somewhat according to different circumstances:

As layers: Leghorns, Hamburgs, Minorcas or Andalusians, Houdans, Brahmas, Spanish, Polish, Dominiques, Game, Cochins, La Fleche. For quality of meat: Game, La Fleche, Dorkings, Crèvecœurs, Houdans, Polish, Brahmas, Dominiques. For size and weight: Brahmas, Cochins, Dorkings, Crèvecœurs, La Fleche, Malays. For hardiness: Leghorns, Brahmas, Dominiques, Cochins, Minorcas or Andalusians, Games. As sitters and mothers: Dorkings, Game, Dumpies, Silkies, Dominiques, Brahmas, Cochins. For a combination of useful qualities generally: Brahmas, Houdans, and Dominiques.

With all the improvements made by pure breeding and selection the general arrangement into classes remains substantially the same. We would now head the list of general-purpose fowls with our favorite, the Plymouth Rock.

EGGS ONE CENT EACH.

One who has had experience says: "We have kept close account of our receipts and expenses the past year and we find, by buying all the feed at retail, the actual cost of producing eggs was one cent each.

"We used Leghorns, Houdans, Langshans and Wyandottes in the test. So, then, all over twelve cents a dozen is clear gain, while all below that figure would be a loss. Again, in making up the sum we found that the hens averaged three eggs apiece during the week. Where one can grow the feed, the cost should not be more than six to eight cents a dozen. By careful attention, so that fowls lay a large percentage of their eggs in the winter, when prices are away up, it can readily be seen there is a good profit in an egg farm.

"Through the winter I feed wheat screenings, corn, an occasional oat bundle, and warm feed once a day. A dish of milk and another of water stood in the house, also a box of dust, ashes and sand; a cabbage or beet hung from the roof where all could reach it. I kept some thirty odd hens through the winter. They commenced laying in January and I sold from eight to ten dozen of eggs a week through most of that month and February. My hen-house, though cheap, was warm and comfortable."

SHOULD EGGS BE AIRED?

One who has had large experience in artificial incubation says: "I have done considerable experimenting as to whether eggs should be aired or not in incubators.

"I have several times tried two machines, filled at the same time with exactly the same kind of eggs, and sitting side by side; one I would air and the other I would never open more than two or three times during the hatch. I always find that in warm weather, when the eggs are not aired, the chicks are sure to come out too soon and not come evenly, sometimes a few coming out as soon as the eighteenth day, while the others would run as late as the twenty-fourth.

"I have also found that by airing occasionally in warm weather the chicks usually go the full time, and come out much more evenly. I do not air every day at any time, but two to four times a week, seldom more than three times, being governed by the heat of the weather.

"Airing in cold weather I have invariably found to produce poor hatches, and I think that the eggs get all that is needed in cold weather during the times the machine has to be opened, testing eggs, etc. As long as the temperature of a hatching-room is below 50° I do not think it necessary to air eggs. At least that has been my experience."

COMFORT OF POULTRY WHILE SHIPPING.

To lessen the suffering of poultry while being shipped, follow these rules: Do not crowd them. Place cups for water at the four corners of the coop, and also midway between. Place boxes of feed by the side of the cups. Put sand and gravel on the floor. Have the coop at least twenty inches high, with a cloth top, and the sides open, so as to protect from the sun as well as provide air and keep the coop cool.

DOWNY FOWLS.

As far back as I have been able to learn, sports with downy plumage have occasionally occurred.

In the year 1887 I raised three chicks that had downy plumage and purchased two that were raised in a neighboring town. This gave me a pair of Rose-Combs and a trio of Single-Comb Downy fowls.

I have been breeding them in separate yards ever since, and have had no return to the ordinary plumage of other fowls. The males in single combs came from pure-bred Plymouth Rocks. The males in rose combs came from the American Dominiques. The female ancestry of the black and light varieties could not be traced.

I am now breeding them in six varieties, both rose and single combs, of three colors, which I have named Dominique Downys, Black Downys and Light Downys. The cut of the hen on page 263 shows the Light Downy.

These fowls are of medium size, the weight being about five and a half for hens and seven for roosters. As chickens they are rapid growers and are never naked or ungainly looking. As fowls, they have, so far, proven themselves to be healthy, hardy and prolific, being especially adapted to a cold climate. They are very docile and easily confined, making splendid mothers. They dress nicely for market at any age, having plump, round bodies and clean legs.

The peculiar characteristic of the Downy fowls is the structure of the feathers, which are entirely devoid of any web, each spray standing out separate and distinct, giving them the appearance of miniature ostrich plumes. The large feathers of the wings and tail show only a trace of web, hence it is impossible for them to fly at all, and being unable to fly, any fence they cannot jump over or crawl through, though but two and a half or three feet high, is sufficient to confine them. The body feathers all being soft and fluffy, are equally as good as goose feathers for bedding, making quite an item in their favor.—W. D. Hills, Odin, Ill.

BROILERS

are young chickens that have been petted and pushed through

the eight to ten weeks of their lives till fat and fit for the frying-pan.

In the Boston market dark-colored eggs are wanted, while they cannot be too white for the New York taste.

DOWNY FOWL.

Bowker Animal Meal is a good remedy for feather-eating according to Wm. Shepard who says: "My hens were picking the feathers from one another, but soon stopped after feeding the meal. They were laying when I commenced feeding it, but it increased the number of eggs."

IMPORTS OF FOREIGN EGGS FOR ELEVEN YEARS.

Year ending June 30.		Dozens.	Value.
Under Free Trade.	1883	15,279,065	$2,667 604
	1884	16,487,204	2,677,630
	1885	16,098,450	2,476,672
	1886	16,092,583	2,173,454
	1887	13,936,054	1,960,896
	1888	15,642,861	2,312,478
	1889	15,918,809	2,418,976
	1890*	15,062,796	2,074,912
Pro-tect'd	1891	8,283,043	1,185,595
	1892	4,188,493	522,240
	1893	3,295,842	392,617

*Protected October 6, 1890, at 5 cents per dozen.

We find during the eight years, from 1883 to 1890, that we imported on an average more than 15,500,000 dozen eggs every year; over 124,515,000 dozen of foreign eggs sold in this country in eight years. What did they cost? More than $18,-770,000, almost $50,000 every week of the eight years sent abroad to foreign farmers, and yet some people in the States seem to think the poultry business is in danger of being overdone. We think not yet a while.

As will be seen from the above table we have each succeeding year since 1889 bought fewer foreign eggs. These have come from Canada and from China for use in California. None from Austria or Belgium, France, Germany, England, Scotland, Italy or the Netherlands.

EXPORTS OF EGGS FOR TEN YEARS.

Year.	Dozens.	Value.
1883	360,028	$75,080
1884	295,484	60,750
1885	240,768	51,832
1886	252,202	46,105
1887	372,772	60,686
1888	419,701	66,724
1889	548,750	75,986
1890	380,884	58,675
1891	363,116	64,259
1892	183,063	32,374
Total	3,416,763	$594,439

The average importing price was fifteen cents per dozen, while the exporting price averaged within a small fraction of seventeen and one-half cents per dozen.

IMPORTS OF EGGS BY COUNTRIES.

During the fiscal year ending June 30, 1889, the United States bought from foreign countries almost 16,000,000 dozen of eggs, for which we paid almost $2,420,000 or more than fifteen cents per dozen. The custom-house returns give the following showing:

From	Dozens.	Dollars.
Austria-Hungary	$ 1,528	882
Belgium	215,164	83,223
China	126,300	6,425
Denmark	74,950	11,899
France	140	24
Germany	73,355	14,119
England	4,914	897
Scotland	4,100	820
Nova Scotia, New Brunswick and Prince Edward Island	3,637,222	481,609
Quebec, Ontario, Manitoba and the Northwest Territory	11,731,864	1,864.020
British Columbia	975	86
Hong Kong	15,219	780
Italy	12,468	2,078
Japan	20	5
Mexico	18,587	2,380
Netherlands	500	70
Cuba	1,503	154
Turkey in Africa	....	5
Total	15,918,809	$2,418,976

WHERE FOREIGN EGGS WERE RECEIVED IN 1889.

The following table, compiled from the official statistics of the Treasury Department at Washington, shows the towns and cities which received foreign eggs during the year ending June 30, 1889:

At	Dozens.	Dollars.
Aroostook, Maine	1,958	$ 277
Bangor, Maine	546,826	68,142
Bath, Maine	385	35
Boston and Charlestown, Mass	1,938,267	270,990
Buffalo Creek, N. Y.	5,740,946	920,096
Cape Vincent, N. Y.	9,400	1,347

Champlain, N. Y........................	829,894	125,603
Corpus Christi, Texas....................	32	4
Cuyahoga, Ohio..........................	600	80
Detroit, Mich............................	487,993	54,314
Genesee, N. Y...........................	735	104
Gloucester, Mass.........................	15,783	2,342
Huron, Mich.............................	11,777	1,418
Key West, Fla...........................	1,503	154
Marblehead, Mass........................	1,729	204
New London, Conn.......................	816	83
New York, N. Y..........................	892,469	63,845
Niagara, N. Y...........................	1,412,963	240,686
Oswegachie, N. Y........................	1,020,293	141,521
Oswego, N. Y............................	25	5
Paso del Norte, Texas, and New Mexico....	18,555	2,376
Passamaquoddy, Maine	1,122,638	138,131
Philadelphia, Pa.........................	....	5
Portland and Falmouth, Maine.............	2,681	328
Portsmouth, N. H........................	164	18
Providence, R. I.........................	30	6
Puget Sound, Wash.......................	975	86
Salem and Beverly, Mass..................	4,184	563
Sandusky, Ohio..........................	720	62
San Francisco, Cal.......................	126,300	6,425
Superior, Mich..........................	7,764	1,162
Vermont................................	2,256,070	377,407
Waldoborough, Maine.....................	8,364	485
Willamette, Ore..........................	10,215	507
All other customs districts................	1,255	211
Total.............................	15,918,809	$2,418,976

HOW TO SELECT A GOOD LAYER.

A writer in Nor'West Farmer and Miller says:

"How many poultrymen can pick out a good laying hen from a strange flock? Not many can do it; yet it can easily be done after a short study of make-up and characteristics.

"There goes a hen with a thick neck, large head, ill-shaped, walks listlessly about, seemingly with no intention or purpose in view. She doesn't care to scratch, but hangs around the hen-house, evidently waiting for her next feed. She gets up late in the morning, and goes to bed early in the evening. That hen may be put down as a very poor layer. The eggs of some of the other hens go to help pay her keep.

"Here comes another hen, She walks briskly, and there is an elasticity in her movements that denotes she has something in view. She is neat and natty in appearance, small head, with a slim neck nicely arched or curved. She forages and scratches all day long, and may be too busy to come for her evening meal. She is at the door in the morning waiting to be let out. She snatches a few mouthfuls of feed, and is soon off to the meadow looking for insects. Before she gets out in the morning she generally deposits her daily egg in the nest, or returns after a short forage. She is neat, clean and tidy, with a brightness and a freshness pleasant to the eye. That is the hen that pays for her feed and gives a good profit all the year round.

"The writer has noticed these traits since boyhood, and knows that they are infallible. By studying these traits, any man may in a few years, by selection, have a fine laying flock of hens."

SPONGIA FOR ROUP.

When roup is in the croup or hoarse breathing stage, spongia may be given with good results. This is a homeopathic remedy and what is known as third potency is generally used. Homeopathic remedies should be used sparingly. Ten drops, or twenty pellets, in a quart of drinking water, placed where all the birds can drink it, is the proper method. Do not use more than ten drops. Use clean vessels, which should be washed and scalded every day. It is considered the best remedy for hoarseness that can be given.

DEFORMED CHICKS.

From Winnipeg comes an inquiry in regard to deformed chicks, Mr. A. F. Preston, of that city, sending us the following letter, which calls attention to a very frequent occurrence:

"We have the hot-water incubator, and in the first hatch the chicks' feet seemed to be weak, and they could not straighten their toes. A great many were too weak to get out of their shells."

It is due to several causes, but usually too high a temperature when the toes are crooked, but if weak and not thrifty for the first twenty-four hours it denotes heat may have been low. We could not well answer the inquiry unless the details of hatching were given.—Poultry Keeper.

CAMPINES.

Since this Belgian breed has lately been admitted to the American Standard by some enthusiast, your readers may be interested in learning that "Campine" is only a French name for what an American would call a Penciled Hamburg; so that in making a place for Campines in the next Standard, it will be describing the same fowl under two different names.

There are six varieties of Campines:

1. Single-Combed Silver.
2. Rose-Combed Silver.
3. Single-Combed Golden.
4. Rose-Combed Golden.
5. Single-Combed White.
6. Short-Legged Single-Combed Silver.

Rose-Combed Whites have been seen also, but are so rare as to be unworthy of notice. Varieties two and four are identical with Standard Silver and Golden-Penciled Hamburgs, and one and three are different only in comb. If the A. P. A. had really desired to make a place for these Belgian fowls, the proper thing would have been to have merely added some single-combed varieties to the Hamburg class. But these Belgian fowls have absolutely no merit to recommend them for exhibition purposes, and their "wonderful laying" is not any better than that of any other Hamburg under the same conditions.—H. P. Clarke, in Poultry Monthly.

LICE IN A NUTSHELL.

1. When chicks droop, and appear sick without cause, especially in summer, look for lice—not the little red mites, but the large gray body lice on the heads and necks.

2. If you find them use a few drops of grease of any kind. A teaspoonful of oil of pennyroyal to a cup of lard is excellent.

3. Look under the wings for the red lice, but use only a few drops of the lard.

4. Never grease the bodies of chicks unless lightly, as grease will often kill them.

5. Never use kerosene on chicks, unless it be a teaspoonful of kerosene to a teacup of lard, as it is irritating.

5. Crude petroleum is always excellent, and serves as a liniment, but mix it with twice its quantity of lard.

7. Keep the dust bath always ready. Use dry dirt or sifted coal ashes. Add Persian insect powder or oil of pennyroyal to the dirt.

8. To rid the house of lice sprinkle coal oil everywhere—floor, walls, roosts, yards, roof, inside and outside, and repeat often.

9. Dust insect powder in the feathers, and be sure it is fresh and good.

10. Put insect powder and tobacco dust in the nests. Never pour grease in the nests. Clean them out every week.

11. Even when no lice may be present use the sprinkler of kerosene at least once a week; and keep the roosts always saturated.

12. No matter how clean things may appear, look for the large lice on the heads, throats and vents.

13. Lice abound both in winter and summer, but more especially in summer.

14. One-half the chicks and young turkeys die from lice. Chicks or turkeys with hens or turkey hens always have lice (either the mites or large lice). Remember that.

15. Always aim to get the solutions or powders into the cracks and crevices.

16. The easiest and best way to whitewash is with a force pump. They are now made to force water from a bucket.

17. When your chicks have bowel disease look for the big lice.

18. No mites need be present where plenty of coal oil and carbolic acid are used.

19. Lice means work. Repeat these precautions and remedies frequently.—P. H. Jacobs in Poultry Keeper.

THE BLACK RUSSIAN.

Since the meager description of the Russians in the chapter on breeds was in type we have obtained the following information. The origin of the breed is imputed to the Cossacks of central or southeastern Russia. A well-known judge asserts that it is an American variety, to which the name "Russian" was arbitrarily given. In plumage the Russians are a glossy greenish-black, very thickly feathered, especially about the head and neck. The comb is double or rose, fitting close to the head. It should have small or no spikes. The wattles are small and nearly hidden by the beard. The breast and body are full and deep, the skin yellow, and the legs dark lead color, shading to yellow. The bottoms of the feet are yellow.

The Russians endure vigorous winter weather, as the comb is small and of very tough consistency, being nearly frost-proof. The hens are acknowledged to be among the best of winter layers by all who know them, and good sitters the year round, although the same hen is not apt to sit more than once in a season. Some do not sit at all. The fowls are of medium size, weighing from six to eight pounds at maturity. They are excellent table fowls. Thus they combine the qualities of utility and beauty, being of stylish carriage and appearance. A fence four feet high is sufficient to confine them. The most serious drawback at present is the difficulty of obtaining non-related blood.

CAPONS OR COCKERELS.

Replying to an article by Mr. Leggett, part of which we quoted in our chapter on capons, about the profitableness of cockerels as compared with that of capons, Mr. J. Frank Tatum, of Maryland, in a recent Country Gentleman says: Mr. Leggett says, "Having looked carefully into the capon business, it suits me best to sell my surplus cockerels as they are." He has looked into it, but he has not given it a fair trial. Until lately I lived in central New Jersey. I raised capons and shipped to commission men in New York City, so they were sold in open market at wholesale. I think the proper test is to raise some of each from the same brood, that is hatched at just the same time; take birds as evenly matched as possible at time of caponizing, then give them exactly the same chance in every way, dress one lot as carefully as the other, kill all the same day, and sell all through the same man on same day. I have always found a difference in price of from six to eight cents a pound, and sometimes more. This fall I sent a lot on for Thanksgiving; this was one of the worst markets ever known in New York. The capons sold for seventeen cents per pound; the chickens for ten cents. All were hatched about May 1. Last March I got twenty-three cents for capons; chicks brought only sixteen cents.

Now, it may not pay Mr. L. to have his cockerels caponized. He raises only pure-bred birds. He may be able to sell at fancy prices for breeding. If he can, it is quite likely that will pay better. It is often stated that capons will weigh more than cockerels of same age with same care. If the cockerels are cared for as Mr. L. advises, I doubt if there will be any difference in weight, but if they have free range, mix with hens, etc., there will be. Up to the time cockerels begin to run hens, there will be no difference; from that on, the capons will gain faster. He says "capons need from fourteen to eighteen months for full growth." I believe that to be so, but don't see why one need keep them until full grown. If they are the right kind and well cared for, they will do to market long before that. As before stated, I sold capons this fall at six and one-half months. They dressed eight pounds each. I never kept a capon full twelve months. In New York market above eight pounds brings top price; if much below that, goes at "slip" prices, usually from two to three cents a pound less. On some markets people don't know what a capon is. On such a market capons may not pay.

Mr. Leggett says: "It is not fair to quote price on extra fine capons and on second-quality cockerels and compare notes." I say it is no more fair to compare extra fine cockerels with second-class capons. He implies that the loss from operation is quite large. I have had experts caponize my chickens; have also caponized both my own and others' cock-

erels.   I am sure, if the chicks are healthy and the operation
is properly done, the loss is so very small that it is not
worth notice.   If one should happen to die, it will be at once,
from bleeding to death, and it is good to eat.   There is no
extra care needed that I know of.   I don't think we need
charge capons with any extra cost, except fee for operation,
usually three cents each (unless it may be best to hire an
expert to dress them).   They must be dry-picked.   Any
careful person can soon learn to pick dry.

Now, say the capons and cockerels each weigh eight
pounds; if the capon brings seven cents a pound more there
is fifty-six cents on each bird.   If we have a small capon or
a slip, it will be four cents a pound or thirty-two cents gain
on each.   This is supposing that the cockerels are cared for
as Mr. L. advises, and all are sold on a capon market at the
same time, but the most chickens are raised by farmers or
others who only raise comparatively few, and think they
can't yard any of their poultry (a mistaken idea, I think).
In this case the difference in favor of the capon will be decidedly more, as he will weigh more than the cockerel.   As
Mr. L. says, the market has much to do with it.   I know
what I say is true for New York market, but on some other
market there may not be so much difference; I am sure, however, there will be quite a difference where people know fully
what a capon is.   Still every one must judge for himself
what js best on his market.

What I have said so far is simply with reference to market value, but unless I had a better market for breeding birds,
or was sure all cockerels could be kept entirely by themselves, I would surely have all cockerels caponized even if it
added nothing to their market value.   Most chicks are raised
with free range.   From the time the males are old enough to
run the hens, the cockerels not only don't do so well themselves, but keep the hens from doing as well as they should.
It is very seldom a true capon will ever crow, fight or notice a
hen.   A slip is usually about as bad as a cockerel.

Tastes differ and I may be mistaken, but I think capons
much better eating.   If they have had a free range, they
certainly are far more tender.   Almost any one will admit

that steer meat is better than bull meat; then why should not capons be better than cocks?

It is often stated that any one can caponize. He can, some way, if careful, but no one without a good deal of practice can do first-class work. A new hand will be sure to make a lot of slips. If careful, he need not kill any. Even an expert will make a slip now and then. The more practice one has had, the fewer slips. There is a great deal of difference, not only between birds of the same breed, but between different breeds. As a rule, Partridge Cochins are easy to do, while Brahmas are hard. The first year I cut any, I cut seventy-four, all my own and largely Light Brahmas. I worked for about an hour on one chick and then gave it up, but cut it all right two or three weeks later. I could probably do the same bird in five minutes now. I killed three (all of which we ate) learning how, and so far as I know, this is all I ever lost from operating. Of the seventy-four the first year, there were, I think, seventeen slips. I therefore say if you can get an expert, do so; but if you can't, do it yourself. Be sure to have the right instruments. The right kind have a hair holder. Any other kind are almost sure either to make slips or kill your bird. Any time before they start to run hens will do; after that, you had better not try it. If they are perfectly healthy and you are careful, you are not likely to lose any; but if they are not healthy, they are apt to die, no matter how careful you are.

FERTILIZATION OF EGGS.

The question is asked: "How soon are eggs fertilized after the male is admitted to the flock?" It is probable that with chickens life is imparted to the germ of the fowl by each specific copulation. The hen is most susceptible immediately upon leaving the nest after laying. It is claimed by some that ten days should elapse before it is safe to count upon the egg being surely fertilized. One copulation fertilizes the whole litter or batch of eggs a turkey hen lays.

As to the number of cocks to have with a given number of hens, too many are as bad as too few. In fact, some claim that too many are worse than too few, for when the number of male birds is out of proportion to the number of females in a flock, the hens are so worried with the attention given that they get disgusted and mad and dodge them entirely. If the flock has free range there may be more hens with one male than when they are kept in yards.

A writer who has given much study and observation to the subject of egg fertilization says: "I used to have the notion that it was a good plan to keep the sexes apart until ten days or so of the time when the eggs were wanted for incubation. That notion is gone. I found by experience that when the roosters were with the hens right along for a month or six weeks before the eggs were wanted for incubation nearly every egg was fertile; while on the other hand, when a strange rooster was put with the hens only ten days or so previous to the time of using the eggs for setting, a large per cent were infertile; therefore, when you have to buy male birds, I advise you to buy early—in the fall if possible—certainly early enough so that they can be with the hens at least a month before you set the eggs. When you raise your breeding stock let the sexes run together right along after the moulting season is over. When your fowls are once mated do not, unless absolutely necessary, change roosters during the hatching season; if obliged to do so, do not, if you want the eggs to hatch true to the new mating, set the eggs laid during the first ten days after the change of male birds."

DO NOT USE HEAVY ROOSTERS FOR LIGHT HENS.

When crosses are made of light and heavy breeds, as for broilers, the male should be of the light breed, as a heavy male is apt to injure the hens of a lighter breed. If the hen is not ruined as a breeder, her wings may be broken or she otherwise so hurt as to render her of no use except for the pot, and sometimes fever sets in before the injury is discovered, in which case she is unfit for table use.

APOPLEXY.

When one of the biggest and best of the flock acts as though it was in a spasm, tries to stand on its head or whirls and runs about as though it was crazy, you may be pretty sure the trouble is apoplexy. Sometimes one of these nice, fat ones will be found dead under the roost. In this case death was very likely caused by apoplexy. Prevention is better than cure, for generally when a bird is discovered going with apoplexy it is so near gone that it dies in spite of efforts to save. One of the symptoms sometimes seen before it is too late to save is drooping wings and a staggering walk. So if a fat fowl is seen going as though drunk, catch and give a dessertspoonful of castor oil and shut up for twenty-four hours where food can not be had, then feed lightly on cooked food for a week. If pretty near gone when you notice it, hold its head under a stream of cold water or plunge it head first into cold water. If it revives give it the treatment advised above. The best preventive of apoplexy is to feed sparingly of corn in hot weather and to give plenty of exercise to the fowls at all seasons of the year.

VERTIGO

is produced by too much blood or too much fat, and causes the fowl to run about as though confused and unable to tell which way to turn. Similar to apoplexy. To cure keep in a cool, shady place, feed sparingly and give occasionally three grains of jalap.

PIP

is supposed to be produced by indigestion, which causes an inflamed mouth and tongue, resulting in a horny scale on the point of the tongue. Some claim this scale should be removed with the point of a sharp penknife and a pinch of powdered chlorate of potash dropped in the fowl's throat and upon the tongue. Others claim the pip should not be forcibly removed, but that the cause, indigestion, should be overcome by administering daily two or three grains of black pepper in fresh butter. We have never had occasion to try either cure.

INDEX.

DIRECTORY.

BREEDERS OF POULTRY.

The following list includes the names of breeders who sell eggs for hatching, or stock for breeding purposes, or both eggs and stock:

ANCONAS.

Geo W Davis, M. D., Pleasantville, Maryland.
Francis A Mortimer, Pottsville, Pennsylvania.

ANDALUSIANS.

A H Cook, P. O. Box 181, Champlain, New York.
M Coleman, P. O. Box 187, New Rochelle, New York.
Ed Hoffman, Glenville, Cuy County, Ohio.
Sam Spear. Eureka, Illinois.
J Dilworth, 170 King street East, Toronto, Ontario.

ARGONAUTS.

H S Babcock, Providence, Rhode Island.

ASTRACHANS—BLACK.

Geo W Davis, M. D., Pleasantville, Maryland.
W A Williams, Rome City, Indiana.

BANTAMS—BLACK AFRICAN.

E M Crossman, Batavia, Illinois.
R A Chester, Box 612, Sarnia, Ontario.
Henry Muehlenfeld, 914 Kentucky ave., Quincy, Illinois.

BANTAMS—COCHIN OR PEKIN—BLACK.

J F Knox, 162 Crescent avenue, Buffalo, New York.

BANTAMS—COCHIN OR PEKIN—BUFF.

D H Lowry, Brussels, Ontario.
Theo Sternberg, Ellsworth, Kansas.
Henry Muehlenfeld, Quincy, Illinois.
Thos Lund, Landenburgh, Pennsylvania.

BANTAMS—COCHIN OR PEKIN—WHITE.

Whitney Bros, Gouverneur, New York.
Harry J Streuber, Erie, Pennsylvania.

BANTAMS—BOOTED WHITE.

C E Rockenstyre, Albany, New York.
Oldrieve & Nicol, Kingston, Ontario.

BANTAMS—COO-COO.

Middleton & Heys, Waverly, Massachusetts.

BANTAMS—GAME.

Bernard Mohan, Reading, Pennsylvania.
E R Spaulding, Jaffery, New Hampshire.
H H Krier, Owatonna, Minnesota.
S E Wurst, Elyria, Ohio.
Wm Barber, Toronto, Ontario.

BANTAMS—JAPANESE.

Richard Oke, London, Ontario.
W C Geyer & Co., Norwich, Ohio.
Kinter & Co., Box 22, Dillsburgh, Pennsylvania.
Edward Andrews, Box 551, Adams, Massachusetts.
J H Matthews, Tarkio, Missouri.

BANTAMS—JAPANESE—WHITE.

Whitney Bros., Gouverneur, New York.

BANTAMS—POLISH—WHITE-CRESTED WHITE.

W H Lewis, Huntington, New York.
Chas L Seeley, Lock Box 4, Afton, New York.
Geo Kain, Galt, Ontario.
J R Brabazon, Delavan, Wisconsin.

BANTAMS—POLISH—BEARDED WHITE.

F B Zimmer & Co., Lock Box 207, Gloversville, New York

BANTAMS—ROSE-COMBED—BLACK.

Frank D Lewis & Bro., Amsterdam, New York.
Frank P Quinby, Box C, White Plains, New York.
Richard Oke, London, Ontario.

BANTAMS—ROSE-COMBED—WHITE.

J C Hilke, Box 774, Canajoharie, New York.

BANTAMS—SEBRIGHT—GOLDEN.

FRANK D. LEWIS & BRO., Amsterdam, N. Y.

Phil Williams, Taunton, Massachusetts.
A C Ewing, Mt. Cory, Ohio.
Geo Brisbin & Co., Clyde, New York.
Jno Woolley, Sec., Streator, Illinois.
J G Carter, Box 167, Picton, Ontario.

BANTAMS—SEBRIGHT—SILVER.

FRANK D. LEWIS & BRO., Amsterdam, N. Y.

C H Proper, Summit, New York.
S J Titus, North Norwich, New York.
E M Hunt, Shelton, Connecticut.
W McNeil, 774 Waterloo street, London, Ontario.

BRAHMAS—DARK.

W A Fuller, Fultonville, New York.
Jas McLaren, Stephens street, Owen Sound, Ontario.
Edgemount Poultry Yards, Reading, Pennsylvania.
Chas Gammerdinger, Columbus, Ohio.
W N Boyles, Greensburg, Indiana.
Levi Keys, Vermillion Grove, Illinois.

BRAHMAS—LIGHT.

Casper Dice, Roca, Nebraska.

J H Warner, Niskayuna, New York.
Mrs E M Haskell, Exeter, Illinois.
L H Morse, Newark New York.
Mrs B F Scott, Burlington, Kansas.
Seeley Bros., Lansing. Michigan.
Roy B Clark, Chatham, New York.
A G West, Fayetteville, Walworth County, Wisconsin.
Haycock & Kent, Kingston, Ontario.

COCHINS—BLACK.

Theo Sternberg, Ellsworth, Kansas.
R P Thompson, Box 48, Patterson, New Jersey.
R G Buffinton, Fall River, Massachusetts.
W McNeil, 774 Waterloo street, London, Ontario.

COCHINS—BUFF.

Dr. H F Ballard, Chenoa, Illinois.
J A Secord, Searsburg, Schuyler County, New York.
F C Hare, Whitby, Ontario.
J A Wilson, Amesbury, Massachusetts.
E T Blood, Kent, Ohio.
E B Easter, 1440 Grand River avenue, Detroit, Michigan.
J W Russell, Vermillion, South Dakota.

COCHINS—PARTRIDGE.

John and P F Spahr, Carlisle, Pennsylvania.
W Lovell, Box 357, Galt, Ontario.
Geo Neutse, Coshocton, Ohio.
J S Wells, Greenport, Suffolk County, New York.
John Irons, Stronghurst, Illinois.
G D Holden, Owatonna, Minnesota.
J V Gilbert, Muncie, Indiana.

COCHINS—WHITE.

J K Holmes, South Schodack, New York.
E C Stewart, Franklin, New York.
C C Shoemaker, Freeport, Illinois.
Frytown Poultry Farm and Kennels, Hannibal, Missouri.
F R Webber, Box 268, Guelph, Ontario.

CREEPERS.

T T Jones, Prospect, Ohio

CREVECOEURS.

F H Dolbear. Bower's Corners, New York.
C A Sharp & Company, Lockport. New York.
Standard Poultry Club, Albion, Illinois.
Richard Oke, London, Ontario.
Chas. Gammerdinger, Columbus, Ohio.

DOMINIQUES—AMERICAN.

R W Roberts, Camroden, New York.
John B Avery, Stittville, Oneida County, New York.
Thos. H Crowder, Bethany, Illinois.
D H Gile, Darien, Wisconsin.
J M Wise, Freeport, Illinois.

DORKINGS—COLORED.

Henry Hales, Ridgewood, New Jersey.

John Lawrie, Malvern P. O., Ontario.
G F Davis & Co., Dyer, Indiana.

DORKINGS—SILVER GRAY.

J H Pitney, Eagle Bridge, New York.
A J Smith, New Millport, Pennsysvania.
W Westfall, Sayre, Pennsylvania.
Allen Bros., Newcastle, Ontario.
Albion Poultry Yards, Albion, Illinois.

DORKINGS—WHITE.

Arthur L Gardner, Manager, Vermillion, New York.
Beeler's White Stock Farm, Clifton, Indiana.
Freeman & Button, Cottons, New York.
Mrs D B Swift, Shelbina, Missouri.

DOWNY.

W D Hills, Odin, Illinois.

ERMINETTES.

M Watson, South New Haven, New York.

FRIZZLES.

C E Rockenstyre, Albany, New York.
L Rottman, Benton, Ohio.

GAMES—BLACK.

Fowler & Lloyd, Williamstown, Pennylvania.

GAMES—BLACK-BREASTED RED.

J L Corcoran, Stratford, Ontario.
N Bentley, Lock Box 15, Conewango Valley, New York.
Wesley Lanius, Greenburg, Indiana.
W M Clarke, Brookfield, New York.
W R Collie, Delavan, Wisconsin.
W E Walden, Watervliet, Michigan.

GAMES—BLACK RED.

Larkin W Farrar, Buckfield, Maine.
E T Blood, Kent, Ohio.
Swan & Duffield, Wingham, Ontario.

GAMES—BROWN RED.

Geo N Thomas, Trenton, New Jersey.
Foreman Bros., Collingwood, Ontario.

GAMES—DUCKWING—GOLDEN.

Chas E Rogers, New Market, New Jersey.
D A Lowry, Brussels, Ontario.

GAMES—DUCKWING—SILVER.

C H Leach, 11 Montgomery street, Gloversville, New York.
Mrs G D Smith, Preston, Ontario.
Jas Yount, Freeport, Illinois.
R Sutton, Richland Center, Wisconsin.

GAMES—HEATHWOOD.

J Oliver & Son, Charlotteville, New York.
A Oliver, Summit, New York.
John M Jacobs, Box 13, Lansdale, Pennsylvania.

GAMES—INDIAN.

M E Savoss, Edison Park, Illinois.
Robert A Colt, Manager, Pittsfield, Massachusetts.
Dr Otto Von Long, Salem, New Jersey.
D G Davies, 91 Grange avenue, Toronto, Ontario.
John Bauscher, Freeport, Illinois.
A M Bowman, Salem, Virginia.
G Strange, Betzer, Michigan.
E A Haslett, Atchison, Kansas.

GAMES—INDIAN—ASEEL.

H P Clarke, Indianapolis, Indiana.
C H Sharp & Co., Lockport, New York.

GAMES—INDIAN—CORNISH.

C D Smith, Fort Plain, New York.
Mrs Bettie Bates, Hale, Missouri.
Mrs Anna Pollard, Washington, Iowa.
S A Forquer, Hillsdale, Michigan.
J G Bickwell, 314 Vermont street, Buffalo, New York.

GAMES—INDIAN—WHITE.

A W Cone, Painesville, Ohio.
C H Sharp & Co., Lockport, New York.
Abbot Bros., Thuxton, Hingham, Norfolk, England.
Robert L. Shanks, Greenwich, New York.

GAMES—MALAY'S—BLACK-BREASTED RED.

L Rottman, Benton, Ohio.

GAMES—MALAY—WHITE.

Carpenter & Andrews, Box 551, Adams, Massachusetts.
Abbot Bros., Thuxton, Hingham, Norfolk, England.

GAMES—MUFF.

John Dame, Box 163, Wheaton, Illinois.
Larkin W Farrar, Buckfield, Maine.

GAMES—PIT.

Robert A Colt, Manager, Pittsfield, Massachusetts.
M O'Brien, Sherburne, New York.
C Trushel, New Castle, Pennsylvania.
E D Castleton, Washington C H, Ohio.
R C Sikes, Andrew Chapel, Tennessee.

GAMES—RED PYLE.

A F Peirce, Winchester, New Hampshire.
W Barber & Company, Toronto, Ontario.
Geo. N Thomas, Trenton, New Jersey.

GAMES—SUMATRAS—BLACK.

Chas. F Reed, Winchester, New Hampshire.
John and P F Spahr, Carlisle, Pennsylvania.
C C Shoemaker, Freeport, Illinois.
Chas. Gammerdinger, Columbus, Ohio.

GAMES—SUMATRAS—MOTTLED.

W A Barhite, West Salamanca, New York.

GAMES—WHITE.

John J Hall, Clyde, Ohio.

HAMBURGS—BLACK.

H S Frederick, Lititz, Pennsylvania.
Kent & Bennett, Auburn, New York.
A G Brown, Watford, Ontario.
J R Brabazon, Delavan, Wisconsin.

HAMBURGS—GOLDEN PENCILED.

F A Stuart, Marshall, Michigan.
F H Dolbear, Bowens Corners, New York.
B A Ferris & Company, Auburn, New York.
Richard Oke, London, Ontario.

HAMBURGS—GOLDEN SPANGLED.

Julius Frank, 886 Bowery, Akron, Ohio.
Noble, Mick & Company, Ionia, Michigan.
Kansteiner Brothers, St. Charles, Missouri.
Alexander Brown, Allandale, Ontario.

HAMBURGS—SILVER PENCILED.

Arthur L Gardner, Vermillion, New York.
J H Boomer, Atkinson, Illinois.
A Bogue, London, Ontario.

HAMBURGS—SILVER SPANGLED.

Nelson S Haskell, Auburn, New York.
C L Clark, Ashton, Illinois.
Mrs P L Reitz, Pansy, Jefferson County, Pennsylvania
Max A Christopher, Warrensburg, Missouri.
C D Smith & Sons, St. Charles, Minnesota.
J M Greyerbiehl, Guelph, Ontario.

HAMBURGS—WHITE.

J B Foster, Allegheny, Pennsylvania.
A W Hillis, Davenport, New York.
Julius Frank, 837 Bowery, Akron, Ohio.

HOUDANS.

**G. A. HOBART, Chittenango, Madison County, N. Y. 96
point cock at head of pen. Eggs, $1.00.**

E J Mason, Gloversville, New York.
C Stockwell, London, Ontario.
Geo Hobart, Chittenango, New York.
J W Miller Co., Freeport, Illnois.
Yorgey & Rich, Horicon, Wisconsin.
S B Mills, Ames, Iowa.

JAVAS—BLACK.

C Hammerschmidt, South Buffalo, New York.
L W Van Winkle, Camden, New York.
J D Robertson, Box 164, Guelph, Ontario.
Frank Doty, Middletown, Ohio.

JAVAS—MOTTLED.

Jos D Hollinger, Mastersonville, Pennsylvania.
J R Brabazon, Delavan, Wisconsin.

JAVAS—WHITE.

F R Terwilliger, Elmira, New York.
N S Perkins, Fairport, New York.
Herbert Hodgson, Albion, Illinois.
A L Smith, Princeton, Indiana.

JERSEY BLUE.

John Bauscher, Jr., Freeport, Illinois.

LA FLECHE.

Richard Oke, London, Ontario.
Wm Smith, Fairfield Plains, Ontario.

LANGSHANS—BLACK.

H P Nestler, King, Indiana.
Dr F M Robinson, Box 10, Pawling, New York.
Frank Marley, Fremont, Washington.
H G Keesling, San Jose, California.
C J Eisele, Guelph, Ontario.
C M Powers, Kent, Ohio.
C C Harper. Mt Carmel. Illinois.
Henry Mansfield, Rockland, Massachusetts.
Gray & Son, Nashville, Tennessee.

LANGSHANS—BLUE.

Col Jos. Leffel, Springfield, Ohio.

LANGSHANS—MOTTLED.

H G Keesling, San Jose, California.

LANGSHANS—WHITE.

M A Skinner, Geneva, Illinois.
H S Keesling, San Jose, California.
M F Delano, Falmouth, Massachusetts.
Paul S Millspaugh, Ithaca, New York.
E McCormick, Newmarket, Ontario.
Mrs M A Smith, Gilman, Iowa.

LEGHORNS—BLACK.

H L Bean, Spencer, Massachusetts.
Elmer E Homan, Yaphank, Long Island, New York.
Oak Lawn Poultry Farm, Ypslianti, Michigan.
R E Haeger, Algonquin, Illinois.
E D Frick, Colorado Springs, Colorado.
Wm. Patterson, Box 852, North Coaticook, Quebec.

LEGHORNS—BLUE.

Col. J Leffel, Springfield, Ohio.

LEGHORNS—BROWN.

EUGENE KEITH, Nelson, N. Y., Single Comb Brown Leghorns and White Plymouth Rocks.

J H Kaufman, Gardner, Illinois.
Requa Bros., Highland, Mills, New York.
W H McCartney, Bethany, Ontario.
M W Warren & Company. Lancaster, Pennsylvania.
Frank Marley. Fremont, Washington.
J A Bailey, Fourteenth and Stout streets, Denver, Colorado.
W Boatright, Matvein, Arkansas.

A E Warner, Lincoln, Virginia.
H L Manners, Sycamore, Illinois.

LEGHORNS—BUFF.

JOHN BAUSCHER, Jr., Freeport, Ill., Buff Leghorn and twenty-eight other leading Varieties; send two stamps for catalogue.

L P Harris, 315 West Thirty-second street, Lincoln, Nebraska.
J M McNeil, Springfield, Ohio.
Geo. W Randolph, Box 725, Palmyra, New York.
Mrs M J Cohes, Sigourney, Iowa.
Aug. D Arnold, Dillsburg, York County, Pennsylvania.
C Houghtaling, Sharon Valley, Connecticut.
R Taylor, 1174 St. Denis street, Montreal, Quebec.
Mrs Lester Kay, Burley Manor, Ringwood, England.

LEGHORNS—DOMINIQUE.

Geo. Bennett, Binghamton, New York.
F R Terwilliger, Elmira, New York.
F A Stuart, Marshall, Michigan

LEGHORNS—DUCKWING.

E Westcott, South Framingham, Massachusetts.

LEGHORNS—WHITE.

E H Hunt, De Kalb, Illinois.
Will C House, Fort Plain, New York.
L C Bryce, Petaluma, California.
F C Cole, Athens, Pennsylvania.
D C Trew, Lindsay, Ontario.
Harris Brothers, Elmira, New York.
J J Fleck, Tiffin, Ohio.
Glass Brothers, Houston, Texas.

MINORCAS—BLACK.

Merwin A Bartlett, Canton, Ohio.
F E Becker, Vine Valley, New York.
Geo. H Northup, Raceville, New York.
M S Kellogg, Hamilton, Missouri.
A D Tollefson, Sioux City, Iowa.
Geo. G Mc Cormick, London, Ontario.

MINORCAS—WHITE.

T F Edwards, Osceola, Tioga County, Pennsylvania.
Chas W Jerome, Box 102, Fabius, New York.
Thomas A Duff, 267 Landsdowne avenue, Toronto, Canada.
T B Knight, Fox Lake, Wisconsin.

PHOENIX—LONG-TAILED.

C S Jackson, International Bridge, Canada.
E H Weiss, Akron, Ohio.

PLYMOUTH ROCKS—BARRED—PEA-COMBED.

J H Blake, Canajoharie, New York.
F H Davey, Minisink, New York.
John Bley, Vail, Iowa.
Geo R Simmons, Box 620, Salida, Colorado.

PLYMOUTH ROCKS—BARRED—SINGLE-COMBED.

A. W. BRAYTON, Mt. Morris, Ill.

J Bennett, 189 Bathurst street, Toronto, Ontario.
J A Willis, Auburn, New York.
Pure-Bred Poultry Company, Mount Morris, Illinois.
Geo P Moore, St Johnsbury, Vermont.
J S Manning, Columbus, Wisconsin.
H A Kuhns, Atlanta, Georgia.
E Lee, Marshalltown, Iowa.
H S Arnold, Lanark, Illinois.
John McDermott, Cramer, Illinois.

PLYMOUTH ROCKS—BUFF.

J C Hilke, Box 774, Canajoharie, New York.
S C Woolverton, Clyde, Ohio.
J D Wilson, Worcester, New York.

PLYMOUTH ROCKS—WHITE.

E S Lamberson, Frankfort, New York.
F P Rogers, West Chester, Pennsylvania.
W H Wight, Hudson, New York.
F Rettig, DeKalb, Illinois.
W S Grigsby, Lena, Illinois.
John Olson, Meckling, Clay County, South Dakota.
L G Pequegnat, New Hamburg, Ontario.

POLISH—BEARDED GOLDEN.

Gulliford & Son, Akron, Ohio.
John M Parker, Independence, Iowa.
John & P F Spahr, Carlisle, Pennsylvania.

POLISH—BEARDED SILVER.

H W Heath, Piermont, New Hampshire.
J R Brabazon, Delavan, Wisconsin.

POLISH—BEARDED WHITE.

D J Pearce, Hamilton, Ontario.
John and P F Spahr, Carlisle, Pennsylvania.

POLISH—BUFF-LACED.

Gulliford & Son, Akron, Ohio.
C S Jackson, International Bridge, Canada.

POLISH—GOLDEN.

M M Mowry, Leicester, Massachusetts.
O'Brien & Colwell, Paris Station, Ontario.
Albion Poultry Yards, Albion, Illinois.
Tom Pierce, Liberty, Union County, Indiana.

POLISH—SILVER.

M M Mowry, Leicester, Massachusetts.
Chas McClave, New London, Ohio.
W R Knight, Box 144, Bowmanville, Ontario.

POLISH—SILVER CRESTED.

W H Lewis, P. O. Box 191, Huntington, New York.

POLISH—WHITE.

W H Lewis, P. O. Box 191, Huntington, New York.

A W Hillis, Davenport, New York.

POLISH—WHITE-CRESTED BLACK.

John F Tallinger, Rochester, New York.
Chas R Carlyle, 52 Connecticut avenue, New London, Connecticut.
A F Herbert, Ionia, Michigan.
Jas Elliott, Bement, Illinois.
W O Bowen, National City, San Diego County, California.
O'Brien & Colwell, Paris Station, Ontario.

REDCAPS.

Phil Williams, Taunton, Massachusetts.
J D Studley, Gowanda, New York.
F A Brown, Port Hope, Ontario.
Howard Miller, Cedarville, New York.
H Baxter, Webster City, Iowa.
A B Pomeroy, Kalamazoo, Michigan.
C C Shoemaker, Freeport, Illinois.

RUMPLESS.

C E Rockenstyre, Albany, New York.

SHERWOODS.

L B Drake, Sheldrake, New York.
W Atlee Burpee, Philadelphia, Pennsylvania.
L C Hoss, Kokomo, Indiana.

SICILIANS.

O D Reese, Old Zionsville, Pennsylvania.

SILKIES.

C S Jackson, International Bridge, Ontario.
C E Rockenstyre, Albany, New York.

SPANISH—WHITE-FACED BLACK.

C H Sheres, Clarksburg, Ontario.
E R Gregory, Edmeston, New York.
M F Lyke, New Baltimore Station, Greene County, New York.
Thos M Skinner, Denver, Colorado.
Mrs B G Mackey, Clarksville, Missouri.
John Bennett, Sunman, Ripley County, Indiana.

SULTANS.

Richard Oke, London, Ontario.

WHITE WONDER.

H W Heath, Piermont, New Hampshire.
Henry O Thomason, Storm Lake, Iowa.
F E McMennamy, Henry, Illinois.

WYANDOTTES—BLACK.

Willow Farm Poultry Yards, West Troy, New York.
J W McNeil, Springfield, Ohio.
Standard Poultry Club, Albion, Illinois.
F W Clemans, Jr., Mechanicsburg, Ohio.

WYANDOTTES—BUFF.

F S Mattison, South Shaftsburg, Vermont.
E B Thompson, Amenia, New York.
E O Therin, Vail, Iowa.

WYANDOTTES—GOLDEN.

G W Felton, Barre, Massachusetts.
Jos McKeen, Omro, Wisconsin.
H D Mason & Sons, P. O. Box 75, Fabius, New York.
F W Eby, Shannon, Illinois.
Gray Brothers, Sweetland, Iowa.
T H Scott, St. Thomas, Ontario.

WYANDOTTES—SILVER.

W N Bartram, Akron, New York.
M B Hague, Inglewood, Ontario.
Edwin M Wilson, Babylon, New York.
B E Rogers, Lake Bluff, Illinois.
E S Johnson, Ash Creek, Minnesota.
J R Brabazon, Delavan, Wisconsin.
C E Fickes, South Sioux Falls, South Dakota.

WYANDOTTES—WHITE.

Knapp Bros, Box 503, Fabius, New York.
John Eklund & Co., Busti, New York.
T M Campbell, Darlington, Indiana.
Selbig, Horning & Co., Lena, Illinois.
Aftondale Poultry Farm, 685 East Fourth St., St. Paul, Minnesota.
Wm Langdon, Port Hope, Ontario.

BREEDERS OF TURKEYS.

BLACK.

E H Cook, Union, McHenry County, Illinois.
J R Brabazon, Delavan, Wisconsin.

BRONZE.

Mrs J C Plumb, Milton, Wisconsin.
C D Smith, Fort Plain, New York.
Mrs Jas R Gooding, Frankton, Madison County, Indiana.
J N Brown, Winamac, Indiana.
Jas B Dorr, Little Falls, New York.
F M Munger, De Kalb, Illinois.
P Pirie, Banner, Ontario.

NARRAGANSETT.

D C Hoff, Jr., Centerville, New York.
C C Paine, South Randolph, Vermont.
J R Brabazon, Delavan, Wisconsin.

SLATE.

Chas McCleve, New London, Ohio.
S E Wurst, Elyria, Ohio.

WHITE.

J H Thompson, Jr., Lanesboro, Minnesota.
J G Ridpey, Rainsboro, Ohio.
E S Appelget, Lock Box 35, Hightstown, New Jersey.
Geo W Emerick, Sumner, Illinois.
Mrs Jno Searles, Jr., Hackleman, Indiana.

WHITE MAMMOTH.

W Atlee Burpee & Co, Philadelphia, Pennsylvania.

W K Laughlin. Ft. Dodge. Iowa. (Birds only).
H A Overton, Knoxville, Iowa.
Valley View Poultry Farm, Belleville, Pennsylvania.
Rhode Island Experiment Station, Kingston, Rhode Island.

BREEDERS OF DUCKS.

AYLESBURY.

J F Hiller, Hartwick Seminary, New York.
G R Baxter, Hillsdale, Michigan.
Chas R McClave, New London, Ohio.
Enoch Parr, Harristown, Indiana.

BLACK EAST INDIAN.

Chas McClave, New London, Ohio.

CALL—GRAY.

F J Marshall, Middletown, Ohio.
S E Wurst, Elyria, Ohio.

CALL—WHITE

A W Hillis, Davenport, New York.
Chas. McClave, New London, Ohio.

CAYUGA.

Wm. P Leggett, Salt Point, New York.
W A Fowler, Vernon, Illinois.
S D Mandeville, Sidney, Illinois.
W Shallenberger, Pleasantville, Ohio.

CRESTED WHITE.

Rucker Bros., Literberry, Illinois.
J R Brabazon, Delavan, Wisconsin.

PEKIN.

Jas Rankin, South Easton, Massachusetts.
J R Roy, Coaticook, Quebec.
W A Graham, Smyrna, New York.
J W Bowlus, Williamsport, Indiana.
Mrs May Taylor. Hale, Carroll County, Missouri.
C H Newman, Lisbon, Illinois.
Frank Marley, Fremont, Washington.

ROUEN.

John Bocker, Seneca Falls, New York.
Chas Hopper, Ashley, Ohio.
J C Bogardus, Jr., Knox. New York.
Jas H McKee, Norwich. Ontario.
John Winter, Box 307. Mendota, Illinois.
B H Westlake. Sycamore. Illinois.

MUSCOVY—COLORED.

I X L Poultry Farm, Petersham. Massachusetts.
J R Brabazon, Delavan. Wisconsin.

MUSCOVY—WHITE.

W Shallenberger. Pleasantville. Ohio.

F H Dolbear, Bowens Corners, New York.
J R Brabazon, Delavan, Wisconsin.

BREEDERS OF GEESE.

CANADA.

Theo. L Morgan, St. Paul, Minnesota.
Chas. McClave, New London, Ohio.
Wm. Smith, Fairfield Plains, Ontario.

CHINESE—BROWN.

John B Bain, New Concord, Ohio.
J R Brabazon, Delavan, Wisconsin.
J H Houser, Camboo, Ontario.
W A Shafor, Oneonta, New York.

CHINESE—WHITE.

J R Brabazon, Delavan, Wisconsin.
Bush Brothers, Selden, Ohio.
A Thompson, Allan's Corners, Ontario.

EMBDEN.

Chas. McClave, New London, Ohio.
Earl G Roberts, Fort Atkinson, Wisconsin.
Ira Gregory, Fifer, Illinois.

TOULOUSE.

W N Bartram, Akron, New York.
C F Michael, Fremont, Ohio.
C C Shoemaker, Freeport, Illinois.
M A Brown, Delavan, Wisconsin.
Wm Hanly, Knoxville, Tennessee.

WILD.

D P McCracken, Paxton, Illinois.
T L Morgan, 583 Wabasha street, St Paul, Minnesota.

MISCELLANEOUS.

GUINEAS.

Theo Searles, Box 308, Port Chester, New York.
A W Hillis, Davenport, New York.
A G West, Fayetteville, Wisconsin.
Mrs S E Thomas, Columbus City, Iowa.

GUINEAS—WHITE.

Fred S McGillis, Brighton, Ontario.
Mrs Mary Nix, Hamburg, Iowa.
W N Bope, Van Wert, Ohio.
D A Mount, Manager, Prince's Bay, New York.

PEA FOWLS.

A W Hillis, Davenport, New York.
J F Barbee, Millersburg, Kentucky.
J M Powers, Henry, Illinois,
L R Freeman, Charlotte, Michigan.

PHEASANTS.

Dr Edward More, Albany, New York.
Dr J S Niven, London, Ontario.

DEALERS IN POULTRY SUPPLIES.

BANDS.

C H Latham, Box 145, Lancaster, Massachusetts.
J G Bickwell, 314 Vermont street, Buffalo, New York.
Morgan Bates, 118 Adams street, Chicago.
R G Davis, Providence, Rhode Island.

BONE MEAL.

Fitch Fertilizer Works, Bay City, Michigan.
York Chemical Works, York, Pennsylvania.
J H Devins, Utica, New York.

BONE MILLS AND GREEN BONE CUTTERS.

Webster & Hannum, Cazenovia, New York.
F W Mann Company, Milford, Massachusetts.
Wilson Brothers, Easton, Pennsylvania.

BROODERS.

E Barney, Schenectady, New York.
G W Pressey, Hammonton, New Jersey.
Chas Dunham, Sycamore, Illinois.
Jas A Porter, Greenfield, Ohio.
H A Peterson, Benson, Illinois.

CAPONIZING TOOLS.

Geo Pilling & Son, 115 South Eleventh Street, Philadelphia, Pennsylvania.
Geo Q Dow, North Epping, New Hampshire.
Wm H Wigmore, 107 South Eighth street, Philadelphia, Pennsylvania.

CHICKEN BIT.

Wm H Wigmore, Philadelphia, Pennsylvania.

CLOVER CUTTER.

P A Webster, Cazenovia, New York.
Wilson Brothers, Easton, Pennsylvania.

COOPS.

John A Jackson, Winnebago, Illinois.
G W Costellow, Waterborough, York County, Maine.
Buckeye Incubator Co., Springfield, Ohio.
Dean Manufacturing Company, Tonawanda, New York.
Hale & Jones, Shelbyville, Indiana.

CUTS.

Clarence C DePuy, Syracuse, New York.
Chas. Gammerdinger, Columbus, Ohio.

DRINKING FOUNTAINS.

T D Paul, Akron, Ohio.
Geo. H Stahl, Quincy, Illinois.

EGG BASKETS.

Geo T Pitkin, 3488 Rhodes avenue, Chicago, Ill.
Sprague Commission Co., Chicago, Illinois.

EGG CASES.

Maxwell Bros., Chicago, Illinois.

Disbrow Mfg. Co., Rochester, New York.
P H Bolten & Co., 221 South Water street, Chicago, Illinois.
Elliott Paper Box Co., Boston, Massachusetts.

EGG RECORD.

H A Kuhns, Atlanta, Georgia.

FENCES.

McMullen Woven Wire Fence Co., 120 N. Market street, Chicago, Illinois.
Page Wire Fence Co., Walkerville, Ontario.
Eureka Gate Co., Waterloo, Iowa.
Hulbert Fence & Wire Co., 904 Olive street, St. Louis, Missouri.

FOODS.

Smiths & Romaine, 109 Murray street, New York City, New York.
Bowker Co., 43 Chatham street, Boston, Massachusetts.
Geo H Stahl, Quincy, Illinois.
Chas Gammerdinger, Columbus, Ohio.
L B Lord, Burlington, Vermont.
Gordon Food Co., Coatesville, Pennsylvania.
I S Johnson, Boston, Massachusetts.
The Peels Food Co., Brattleboro, Vermont.

GAPES EXTERMINATOR.

Wm H Wigmore, Philadelphia, Pennsylvania.

GRIT.

Orrs Mills Poultry Yards, Orrs Mills, New York.

HEATERS.

Wood & Paige, 13 Sewall street, Lynn, Massachusetts.

INCUBATORS.

Reliable Incubator and Brooder Company, 217 North Third street, Quincy, Illinois.
George H Stahl, Quincy, Illinois.
A F Williams, Bristol, Connecticut.
G S Singer, Cardington, Ohio.
Jas Rankin, South Easton, Massachusetts.
Hearson & Company, 447 Grove street, Jersey City, New Jersey.
~~~~ State Incubator Company, Homer City, Pennsylvania.
Prairie ~~~ ~~~~~~~~~ Pennsylvania.
J L Campbell, West Elizabeth, ~~~~ ~~~~~~ ~~~esburg, New Jer-
Pineland Incubator and Brooder Company, ~~~~~~ sey.
Geo Ertel & Company Quincy, Illinois.
Buckeye Incubator Company, Springfield, Ohio.
Famous Manufacturing Company, Chicago, Illinois.
H M Sheer & Brother, Quincy, Illinois.
Wood & Paige, 13 Sewall street, Lynn, Massachusetts.
Gerred Incubator Company, 90 De Grassi street, Toronto.
Von Culin Incubator Company, Delaware City, Delaware.
Geo W Murphy & Company, Quincy, Illinois.
Petaluma Incubator Company, Petaluma, California.
H W.Axford, Cottage Grove avenue and Forty-Fifth street, Chicago.

KILLING KNIFE.

Geo P Pilling & Sons, Philadelphia, Pennsylvania.
Wm H Wigmore, Philadelphia, Pennsylvania.

Incubators. By L. O. Fults, Jeffersonville, Ohio. Price twenty-five cents.

The Indian Game. By H. S. Babcock. Price twenty-five cents. Providence, Rhode Island.

Low Cost Poultry House, By J Wallace Darrow. Price twenty-five cents. The Fanciers Review. Chatham, New York.

Management of Young Chicks by P H Jacobs, Price twenty-five cents. Poultry Keeper Co., Parkesburg, Pennsylvania.

128-Page Poultry Book. Price twelve cents. Clarence De Puy, Syracuse, New York.

Philosophy of Judging. A new book by Messrs. I K Felch and H S Babcock, and illustrated by Mr. J Henry Lee. Explains in language all can understand, the principles upon which the scoring of fowls is founded. Price $1. Ferris Pub. Co., Albany. New York.

Pigeon Queries. By J Wallace Darrow. The Fanciers' Review, Chatham. New York. Price twenty-five cents.

Poultry, by G A McFetridge, Oak Park Stock Farm, Hammonton, New Jersey. Price fifty cents.

Poultry Culture: How to raise, Manage, Mate and Judge Thoroughbred Fowls, by I K Felch. 488 pages. Illustrated. Price, postpaid, $1.50. Donohue, Henneberry & Co., Chicago, Illinois.

The Poultry Doctor. Price fifty cents. Boericke & Tafel, 1011 Arch street, Philadelphia, Pennsylvania.

Poultry for Profit, by P H Jacobs. Price twenty-five cents. Poultry Keeper Co., Parkesburg, Pennsylvania.

Poultry Raising and Artificial Incubation. By J M Stahl & E W Wickey, Champion Manufacturing Co., Quincy, Illinois. Price fifteen cents.

The Practical Poultry Keeper. A complete and standard guide to the management of poultry, whether for domestic use, the markets or exhibition; 252 pages. Price $2. By L Wright, England. Orange Judd Co., New York City.

Practical Turkey Raising. By Fanny Field. Price twenty-five cents. R B Mitchell, Chicago.

Profitable Poultry Farming. By Michael K Boyer. Price twenty-five cents. Oak Park Stock Farm, Hammonton, New Jersey.

Profits in Poultry. Useful and ornamental breeds and their profitable management. Profusely illustrated; 256 pages. Price $1. Published by Orange Judd Co., New York.

Some of Lee's Ideas. By J Henry Lee. Price fifty cents. Oak Park Stock Farm, Hammonton, New Jersey.

Stoddard's Books. Price, twenty-five cents each. H H Stoddard, author and publisher, Hartford, Connecticut. Brown Leghorns. Domestic Water Fowl. How to Feed Fowls. How to Preserve Eggs. How to Raise Poultry on a Large Scale. How to Win Poultry Prizes. Incubation, Natural and Artificial. Light Brahmas. Plymouth Rocks. Poultry Architecture. Poultry Compendium. Poultry Diseases. White Leghorns. Wyandottes.

Treatise on the Game Cock. By F W McDougall, Indianapolis, Indiana.

Wyandottes — Silver, Golden White and Black: Their origin, history, characteristics and standard points; how to mate, judge and rear them for exhibition and commercial purposes; with a chapter on their diseases and treatment. By Joseph Wallace.

Price, fifty cents. Ferris Publishing Company, Albany, New York.

POULTRY PERIODICALS.

The Howard & Wilson Publishing Company, Chicago, Illinois, will send any of the papers and magazines named below, postpaid, at the prices given.

Western Poultry Journal (monthly). Price fifty cents. Cedar Rapids, Iowa.

Poultry Topics (monthly). Price twenty-five cents. Warsaw, Missouri.

Poultry Monthly. Price $1.25. Ferris Publishing Company, Albany, New York.

Southern Fancier. Price fifty cents. G M Downes, Manager. Box 402, Atlanta, Georgia.

Puget Sound Poultry Keeper (monthly). Seattle, Washington.

American Fancier (weekly). Price, $1.50. Blunck & Drivenstedt, Johnstown, New York.

Poultry World. Price, $1.00. H H Stoddard, Hartford, Connecticut.

American Poultry Journal (monthly). Price, $1.00. Morgan Bates, Chicago.

Farm Poultry. Price fifty cents. I S Johnson & Co., Boston, Massachusetts.

Practical Poultryman. Price fifty cents. Branday & Son, Whitney Point, New York.

Western Garden and Poultry Journal. Price fifty cents. Chas N Page, Des Moines, Iowa.

Poultry Herald. Price fifty cents. E A Webb, St Paul, Minnesota.

Fanciers' Review. Price thirty-five cents. J W Darrow, Chatham, New York.

Western Breeder. Price twenty-five cents. Owen & Co. Topeka, Kansas.

Poultry News. Price twenty-five cents. E P Cloud, Kennet Square, Pennsylvania.

Midland Poultry Journal. Price fifty cents. C B Harrington & Co., Kansas City, Missouri.

American Poultry Advocate. Price twenty-five cents (monthly). Clarence C DePuy, Syracuse, New York.

Canadian Poultry Review. Price $1 (monthly). H B Donovan, Toronto, Ontario.

Poultry Chum. Price twenty-five cents. DeKalb, Illinois.

Consolidated Fanciers' Journal. Price, $1. Nashville, Tennessee.

Ohio Poultry Herald. Price, twenty-five cents. Tiffin, Ohio.

Poultry Guide and Friend. Price, twenty-five cents. M K Boyer, Hammonton, New Jersey.

Farm and Fowl. Price, fifty cents. L O Fults, Jeffersonville, Ohio.

Fanciers' Gazette. Price $1. B N Pierce, Indianapolis, Indiana.

Ohio Poultry Journal. Price $1. Robert A Braden, Dayton, Ohio.

Fanciers' Monthly. Price $1. Chas R Harker. San Jose, California.

Poultry and Pets. Price $1. G Rogers, Fort Wayne, Indiana.

Game Fowl Monthly. Price $1. C L Francisco, Sayre, Pennsylvania.

Pigeon Fancier. Price $1. J D Able & Co., Baltimore. Maryland.

Poultry Bulletin. Price $1. Jas E Warner. New York City.

Poultry Keeper. Price fifty cents. Poultry Keeper Co., Parkesburg. Pennsylvania.

Michigan Poultry Breeder. Price fifty cents. Geo S Barnes, Battle Creek, Michigan.

Game Fanciers' Journal. Price fifty cents. Geo S Barnes, Battle Creek. Michigan.

New England Fancier. Price fifty cents. A H Hamilton, Danielsonville, Connecticut.